Lasers in Industry

TECHNOLOGIES ■ APPLICATIONS ■ MARKETS

FIRST EDITION

An article compilation from

)PHOTONICS
MEDIA *photonics.com*

PHOTONICS
MEDIA
PRESS

PHOTONICS MEDIA   photonics.com   PHOTONICS MEDIA PRESS

Group Publisher:	Karen A. Newman
Managing Editor:	Michael D. Wheeler
Associate Managing Editor:	Marcia Stamell
Designer:	Janice R. Tynan
Layout:	Suzanne L. Schmidt
	Janice R. Tynan
Copy Editors:	Mary Beth McMahon
	Carol McKenna

Laurin Publishing Co.

President and CEO:	Thomas F. Laurin
Vice President:	Kristina A. Laurin
Vice President:	Ryan F. Laurin
Vice President:	Erik W. Laurin

Laurin Publishing Co., 100 West Street, Pittsfield, MA 01201
A Photonics Media Publication

Table of Contents

Dates indicate when the article originally appeared in print. Also, author titles and affiliations shown in this compilation appeared with original article publication and have not been updated.

Dates indicate when the article originally appeared in print. Also, author titles and affiliations shown in this compilation appeared with original article publication and have not been updated.

Introduction

For more than half a century, Photonics Media has covered the optics and photonics industry with an unparalleled breadth of content. Every issue of our magazines is a reflection of the continued inroads light-based technologies are making in many industries, and the result is more than just a deep well of material — it is a rich resource for learning and advancing.

With this compilation, we have dipped into those considerable archives and assembled a collection of articles on a subject of great interest to our readers and many others. Technology advancements continue to open new application areas for lasers in many industries, and this collection — *Lasers in Industry* — provides a shop-floor view of lasers in use and a wider look at emerging applications and markets.

This book is for anyone working on, implementing or considering the application of lasers for and in industrial settings for materials processing, quality control and production. It will also serve as an introduction to the subject of industrial lasers for those completely new to the subject. We've done the work of gathering these articles and other resources into one volume and we encourage you to use this compilation as a guide to the current state of lasers in industry, a reference tool and a resource for learning.

Please note that the articles here have not been updated from when they originally appeared in our magazines. With the exception of some minor editing for style and a new page design, the articles appear very much as they did when first published.

The original date and magazine issue of publication for each article can be found in the table of contents. Authors are listed as they were in the original publication, but because they may have moved on to new positions in the interim, no contact information is included.

Photonics Media magazines deliver resources to our readers aimed at helping them do their jobs. This collection has a similar aim: By carefully selecting a range of articles from our extensive archive, we set out to deliver a rich, contemporary resource on lasers in industry. We hope it proves to be a valuable addition to your reference library.

Materials Processing

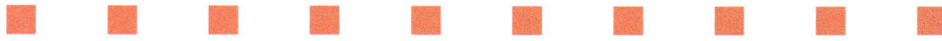

TEA CO$_2$ Lasers Cut Into Excimer Market

Lower costs and higher peak powers are advantages of transversely excited atmosphere carbon dioxide lasers.

BY ROGER SANDWELL, LIGHTMACHINERY INC.

Transversely excited atmosphere carbon dioxide (TEA CO$_2$) lasers are unique in that they offer much shorter pulses and higher peak powers than conventional CO$_2$ lasers. These properties allow TEA CO$_2$ lasers to emulate the better-known — but higher-cost — ultraviolet excimer lasers for some precision processing applications of polymer materials. For manufacturing engineers in the medical device, electronics and printed circuit industries who in the past may have relied on excimer lasers for polymer processing, TEA CO$_2$ lasers can offer lower costs, higher throughput and simpler operation (Figure 1).

Ablation process

Both TEA CO$_2$ and excimer lasers remove material by a process known as ablation. The energy of the laser pulse is deposited in a relatively short time — about 20 ns for an excimer laser and a few microseconds for a TEA CO$_2$ laser — and is

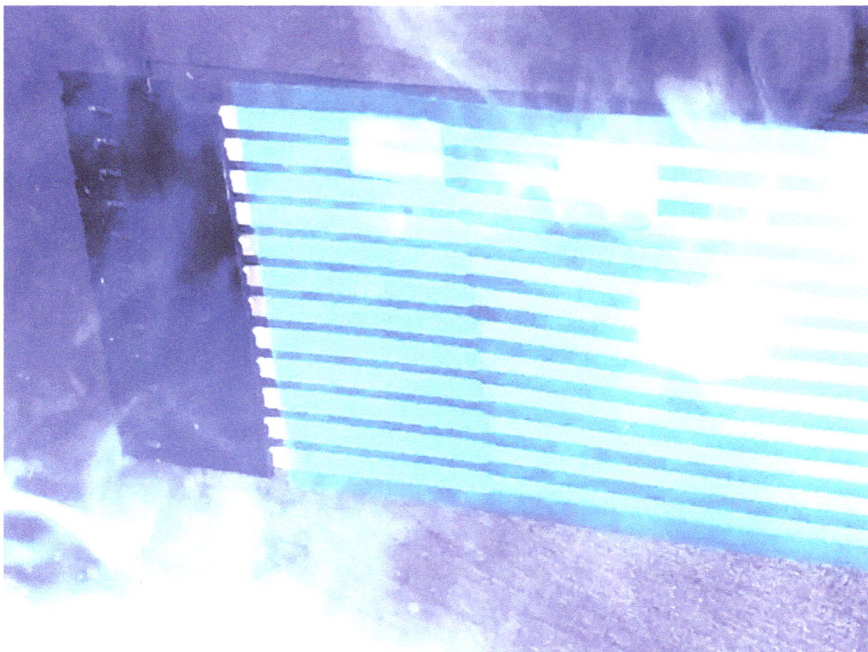

Figure 1. A transversely excited atmosphere carbon dioxide (TEA CO$_2$) laser being used to ablate insulation and expose tracks in flexible ribbon cable. Courtesy of LightMachinery Inc.

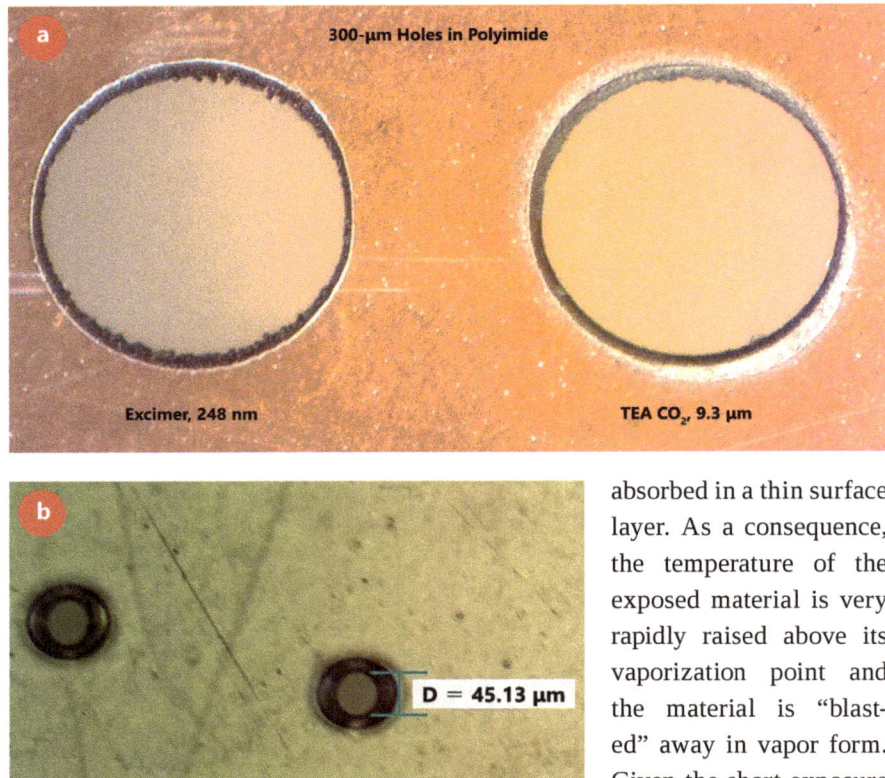

300-µm Holes in Polyimide

Excimer, 248 nm

TEA CO$_2$, 9.3 µm

D = 45.13 µm

Figure 2. A comparison of excimer and TEA CO$_2$ processing of polyimide with 300-µm diameter holes (a) and approximately 50-µm holes (b). The use of a high-quality compound lens was required to achieve these results. Some taper is to be expected in holes with a greater than 1:1 aspect ratio. Courtesy of LightMachinery Applications Laboratory and Optec SA.

absorbed in a thin surface layer. As a consequence, the temperature of the exposed material is very rapidly raised above its vaporization point and the material is "blasted" away in vapor form. Given the short exposure times, very little heat conduction can occur to the surrounding material. Ablated areas tend to have very sharp edges with little or no heat-affected zone (HAZ). This is in contrast to effects observed when processing polymer materials with conventional continuous beam or long-pulse CO$_2$ lasers. With these CO$_2$ lasers, significant thermal damage such as charring, melting and resolidification will occur in surrounding material. Consequently, conventional wisdom says that CO$_2$ lasers cannot be used for fine processing of polymer materials and that excimer lasers are the only option.

However, short-pulse, high peak power TEA CO$_2$ lasers offer an alternative approach for processing of polymer materials with significant benefits of faster throughput, lower capital and operating cost, and simpler implementation when compared to excimer lasers. They can be particularly effective at selective ablation, where a polymer film is removed from an underlying metal layer. The polymer layer may be removed with little HAZ in the surrounding material and no damage to the metal layer, as metal is strongly reflective at these infrared wavelengths.

Ablation characteristics

Typical energy densities (EDs) for ablation of most polymer materials are in the range of 5 to 20 J/cm^2 for CO$_2$ wavelengths. When estimating process speed or throughput, a good number to assume is 20-µm depth of material removal per laser pulse with EDs in the normal range. The maximum depth of material

Excimer vs. TEA CO$_2$		
Feature	**Excimer**	**TEA CO$_2$**
Wavelength	Ultraviolet	Mid-Infrared
Processing Speed	Slow	Fast, ~10 times as effective
Depth Resolution	Excellent	Medium
Minimum Feature Size	1 µm or less	~50 µm
Selective Removal of Polymer Coatings from Metal	Medium	Good
Edge Quality/Low HAZ	Excellent in most cases	Good in many cases
Residue Film	No issue	A problem in some cases
Capital Cost (Dollars per Watt)	High	Around 50% of excimer cost
Operating Costs (Dollars per Hour)	High	Less than 10% of excimer cost
Installation Complexity	Significant	Simple

Table 1. A comparison of excimer and TEA CO$_2$ laser features. Courtesy of LightMachinery Inc.

removal that this author has achieved in a single pulse is about 60 µm at a very high ED around 50 J/cm^2 in a pharmaceutical capsule drilling application.

Similar characteristics generally apply to processing with excimer lasers. Optimum energy density for excimer processing tends to be in the range of 2 to 4 J/cm^2. For most polymer materials, absorption is much stronger at ultraviolet wavelengths than at infrared wavelengths. Absorption depth roughly equates to depth of material removed per pulse. A good rule of thumb for infrared ablation is 20 µm per pulse. For ultraviolet excimer lasers, because of the shorter absorption depth of most materials at ultraviolet wavelengths, the equivalent number is about 0.2 µm per pulse. Consequently, TEA CO$_2$ lasers will give significantly higher rates of material removal and faster process throughput than excimer lasers, but excimer lasers will give better depth resolution. In terms of "raw" removal rate of polymer material, a TEA CO$_2$ laser is at least an order of magnitude more effective than an excimer laser of equivalent power.

Depending on the application, excimer lasers may still be the preferred choice for several reasons:

• **Edge quality:** With their shorter pulse durations and the higher absorption of many materials at ultraviolet wavelengths, edge quality or HAZ in most polymer materials will typically be better when processed with excimer lasers than with TEA CO$_2$ lasers.

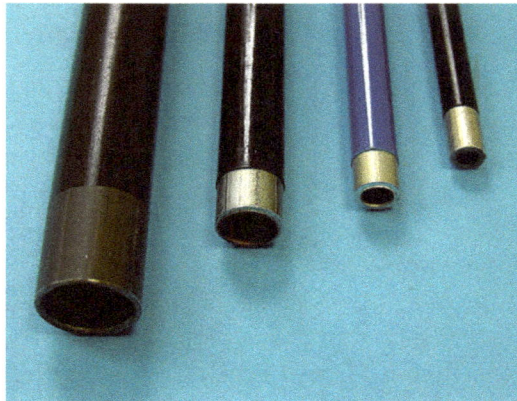

Figure 3. Automobile brake lines stripped with a 300-W TEA CO$_2$ laser. Process time is a few seconds per tube, much faster than could be achieved with an excimer laser. Courtesy of 4JET Technologies GmbH.

• **Feature size:** Because of their longer wavelengths and higher divergence, diffraction effects determine that it is difficult to create feature sizes much smaller than 100 µm with a TEA CO_2 laser and a simple optical beam delivery. With higher-quality optics, feature sizes as small as approximately 50 µm have been demonstrated[1] (Figure 2). Excimer lasers, with their shorter wavelengths, can easily generate feature sizes to a few microns, and in extreme cases (lithographic applications), small fractions of a micron.

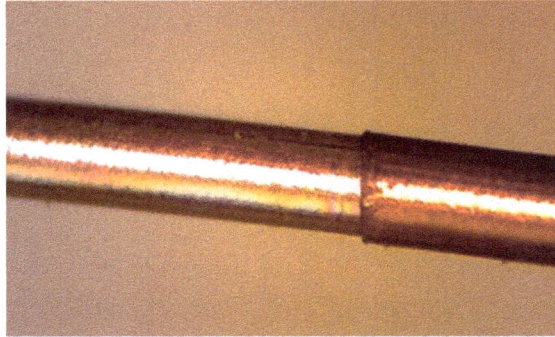

Figure 4. A single-strand single-conductor polyimide-insulated wire stripped with a TEA CO_2 laser. Courtesy of LightMachinery Applications Laboratory

Mask imaging

Excimer lasers and TEA CO_2 lasers share a common characteristic of high beam divergence, which makes it difficult to focus their beam to a small galvo-mirror steered spot as is common with most other types of lasers. Instead, a fixed pattern — round or rectangular hole or some more complex pattern — is created in a thin metal "mask" or stencil; the mask is illuminated by the laser pulse and is imaged with a lens onto the workpiece. Users of excimer lasers will be familiar with this technique.

Wavelength choice

The default wavelength of CO_2 lasers is 10.6 µm, although other wavelengths are available when special resonator optics are fitted to the laser. Brannon and Lankard of IBM[2] were the first to point out in 1986 that the absorption of some polymers, especially polyimide, is much stronger in the 9.3-µm region than at 10.6 µm and that the process quality of such materials is greatly enhanced at the shorter wavelength. All LightMachinery Inc.'s Impact-series lasers supplied for polymer processing, for example, are set for operation at 9.3 µm.

Industrial processing applications

Commercially available industrial TEA CO_2 lasers typically offer average powers, measured as pulse energy by repetition rate, of a few tens to a few hundreds of watts. In this respect, they are comparable to many general-purpose industrial excimer lasers. Applications where TEA CO_2 lasers can be considered as an alternative to excimer lasers include wire stripping, catheter drilling, pad exposure and similar precision polymer ablation in the medical device, electronics and printed circuit industries. Higher-power applications where excimer lasers would not be considered include drilling of controlled-release pharmaceutical capsules and tire-mold cleaning, removal of anticorrosion coatings at the ends of hydraulic brake lines, and paint stripping in the automotive industry (Figure 3).

Wire stripping

A common application for TEA CO_2 lasers is the stripping of fine wires as used in pacemakers and other implantable medical devices, and in magnet assemblies (Figure 4). Many manufacturers of these products tend to believe that only excimer lasers can produce sufficiently clean insulation removal. Although residue issues associated with TEA CO_2 lasers preclude their use in some cases, in other cases they can produce results of acceptable quality at higher throughput — and lower cost when compared to excimer lasers (Figure 5).

Residue issues

There is one issue that may limit the use of TEA CO_2 lasers for polymer ablation in some applications. Because of their longer wavelength and lower absorption, TEA CO_2 lasers typically leave a thin, 1- to 2-μm residual layer of polymer when removing insulation from an underlying metal layer in applications such as wire-stripping and pad exposure. This may or may not be a problem, depending on the follow-on processes. If connections to the wire are to be made by crimping or spot-welding, the residual layer is generally of no concern. Soldering may be more problematical, as is plating to a pad in the printed circuit board (PCB) industry. The residual layer can be removed by a mild chemical wash — generally permanganate — which is a standard process for PCB manufacturers. Medical device customers may be more reluctant to adopt a chemical process. In such cases, dual-laser machines may be considered. These machines include

Figure 5. A typical integrated wire-stripping and cutting machine. Impact-2000 Series TEA CO_2 laser for stripping; Nd:YAG laser for cutting wire. Courtesy of OakRiver Technology Inc.

both a TEA CO_2 laser to obtain the benefits of fast throughput and low operating cost, with a small excimer laser to remove the final micron or two of insulation. In this case the power and duty cycle of the excimer laser can be very low, so the operating cost is acceptable.

Cost comparison

On a dollars-per-watt basis, the capital cost of a TEA CO_2 laser is approximately 50 percent of an excimer laser of equivalent power. The differences in the cost of operation are even more dramatic. A typical 80- to 100-W average power excimer laser in high-duty-cycle operation, such as ink-jet nozzle drilling, will have an operating cost of more than $20 per hour. That takes into account both the consumables cost, such as gas, and the amortized cost of replacement parts, such as the laser tube and other key components with typical lifetimes in the range of 1 to 2 billion pulses. On an equivalent basis, the operating cost of a TEA CO_2 laser will be less than $2 per hour. When the lower cost of TEA CO_2 lasers is combined with the order of magnitude increase in polymer removal efficiency, the case for the TEA CO_2 laser is compelling in all instances where a TEA CO_2 laser provides the required process quality.

Installation and facilities costs will also be lower for TEA CO_2 lasers. Excimer lasers require the use of toxic and corrosive gases, particularly fluorine. This may present a health and safety challenge to some users and at a minimum will require the use of vented gas cabinets and fume exhausts. By comparison, the gases required for TEA CO_2 lasers are relatively benign.

An excimer alternative — not a replacement

An applications engineer who was very familiar with both excimer and TEA CO_2 lasers once commented, "If you've ever used an Impact TEA CO_2 laser, you'd never look at an excimer again." That is an overstatement, because there are some applications where excimer lasers are clearly the preferred choice (Table 1). But in other cases, TEA CO_2 lasers can provide an alternative to excimer lasers with acceptable process quality, lower costs, higher throughput and simpler operation.

Meet the author

Roger Sandwell is the director of sales for LightMachinery Inc. in Nepean, Ontario, Canada.

References

1. Optec SA, Private communication.
2. J.H. Brannon and J.R. Lankard (1986). Pulsed CO_2 laser etching of polyimide. *Appl Phys Lett*, Vol. 48, p. 1226.

Innovations Make Ultrafast Lasers Even Faster

Recent developments in Pockels cells and Faraday isolators are paving the way to increased power and higher repetition rates.

BY MARKUS FEGELEIN, QIOPTIQ PHOTONICS GMBH & CO. KG
AN EXCELITAS TECHNOLOGIES COMPANY

Ultrafast lasers or ultrashort pulse lasers emit pulses in the range of a few femtoseconds up to a few picoseconds. These lasers are essential tools for a variety of applications like laser material processing including cutting, marking, drilling, additive manufacturing and spectroscopic research.

In laser material processing, ultrashort pulses are popular because they only evaporate the material at the laser beam focus position without overheating the surrounding material. Other material processing applications use the high intensity of amplified ultrashort pulses to excite two-photon absorption processes, such as laser cutting of glass, sapphire or cornea — transparent materials that do not absorb the fundamental wavelength but the second harmonic wavelength.

Spectroscopic applications make use of the same two characteristics but in a different way. In analog photography, a light flash or a very short shutter time

Femtosecond lasers are essential in laser material processing. Courtesy of Qioptiq, an Excelitas Technologies company.

reduces film exposure to freeze fast-moving objects. Similarly, ultrashort laser pulses can be considered ultrafast stroboscopes to observe chemical reactions in the femtosecond time frames where they occur.

The broad wavelength spectrum of ultrashort pulses is a key consideration when designing ultrafast lasers. Short pulses are composed of many wavelengths, the more colors contributing to the pulse, the shorter the pulse can be. Therefore every component that is used in an ultrafast laser's beam path must be suitable for operation with a broad spectral range. So not only are mirror and window coatings needed to cover a broad spectral range, but the laser media must also emit and amplify over a broad spectral range (Figure 1).

Dopant determines spectral range

The active laser media is the key element of every laser. For solid-state lasers, this most often consists of a glass or crystalline host to which a dopant is added. This dopant roughly determines the center wavelength of a laser while the combination of dopant and host sets the spectral range of the emission, the laser gain itself and ultimately the length of the laser pulse.

Of all common dopants, titanium, hosted in a sapphire crystal, has the broadest amplification spectrum, and therefore is perfectly suited for the generation of pulses in the range of a few tens of femtoseconds.

For industrial applications, ytterbium (Yb) is commonly used since its pump absorption spectrum is very broad and it efficiently converts pump energy into laser light. When hosted in YAG crystals, the gain is so high that the laser rod can be reduced to a thin disc since the light only needs to travel small distances in Yb:YAG to reach notable amplification[1]. On the other hand, the spectral gain of Yb in YAG is so small that these disk lasers can only emit picosecond pulses. When hosted in vanadate (YVO_4) crystals, the emission spectrum gets much broader, making Yb:YVO_4 the material of choice for femtosecond lasers in the 1030-nm region. Also, the gain of Yb:YVO_4 is considerably smaller compared with Yb:YAG — hence long laser crystals are needed to reach enough amplification.

As a host, crystalline material is preferred because of its typically higher thermal conductivity and damage threshold compared to glass. Glass manufacturing,

Figure 1. Spectral width of a 400-fs pulse emitted from a Yb:YVO Laser. FWHM: full width half medium. Courtesy of Spectra Physics, High Q Laser GmbH, Rankweil, Austria.

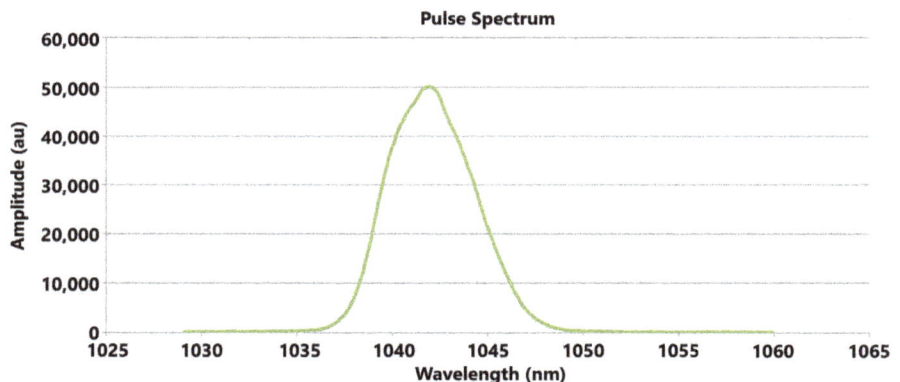

however, is much easier and therefore less expensive and available in large dimensions[2]. Besides the selection of an appropriate laser media mode locking, control of dispersion effects and amplification are important to build a powerful ultrafast laser.

Mode locking

Consider a Ti:sapphire crystal in a laser cavity that is pumped by a continuous wave (CW) laser. If the cavity is correctly aligned, the laser will start oscillating in the free running regime, which means it will oscillate simultaneously and continuously at all resonance frequencies of the cavity within the emission spectrum of the Ti:sapphire crystal. To move the laser from the CW regime to the pulsed regime, all the modes of different frequencies need to be locked on a fixed phase relation between each other (mode locking).

Mode locking can be done in two ways. In active mode locking, the resonator losses are modulated synchronously to the round-trip time of the resonator cavity, for example, by an acousto- or electro-optic modulator. In passive mode locking, resonator losses are modulated by a saturable absorber or a Kerr lens.

For a saturable absorber, the absorption losses decrease with increasing light intensity. Therefore, when the laser starts oscillating continuously, intensity spikes, which are found in the laser noise, have fewer losses than lower or average intensities. After many round-trips, a single pulse remains[3]. The pulse will oscillate in the cavity and as it passes the output coupling mirror, a small portion of the pulse will be emitted. Given a typical resonator length of 1.5 m, the round-trip length is 3 m. Thus the pulse, traveling at the speed of light, will pass the output coupler 100 million times a second, yielding a 100-MHz pulse repetition rate.

Dispersion

Ultrashort pulses take round-trips in the cavity, passing all optical elements within the cavity at least once (for a ring resonator) or twice (for a linear cavity) per round-trip. Since the pulse consists of many colors, it will experience group delay dispersion as typically red light travels faster than blue light through optical media. This will broaden and chirp the pulse, which needs to be compensated by either a pair of prisms or diffraction gratings, or by so-called chirped mirrors for which the penetration depth of red light is deeper than for blue light.

Amplification

Due to the high repetition rate of 80 to 100 MHz, peak intensities of ultrashort pulses emitted by an oscillator are rather small, even if the oscillator yields modest average power. Since many applications, especially in material processing,

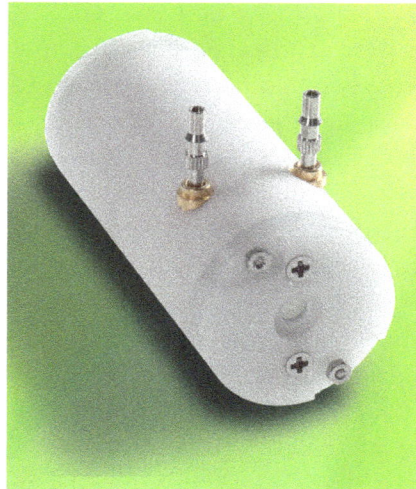

Figure 2. Qioptiq's DBBPC HR Pockels Cell, with two Beta Barium Borate (BBO) crystals for switching frequencies up to 1.3 MHz. Courtesy of Qioptiq, an Excelitas Technologies company.

need much higher peak intensities, it is essential to amplify the ultrashort pulses. This amplification is realized by a master oscillator power amplifier (MOPA) system in which ultrashort pulses of a master oscillator are coupled into an amplifier. Laser media with broad spectral gain often exhibit low gain, so the pulse needs to travel a very long distance through the media to gain considerable amplification.

The most widespread amplifier type for femtosecond pulses is the regenerative amplifier. In this type of amplifier, a Pockels cell picks one pulse out of the 100-MHz pulse train emitted by the seed oscillator into the amplifying resonator where it takes many round-trips through an amplifier crystal. Both end mirrors of the regenerative amplifier are highly reflective to keep the losses per round-trip as low as possible. With each round-trip and each passage through the amplifier crystal, the pulse gains more energy. For Yb:YVO or Nd:Glass the pulse takes approximately 100 round-trips until it is coupled or dumped out again.

Ultrashort pulses can reach very high peak powers relatively quickly. To avoid damage to optical elements, pulses are stretched before they are coupled into the amplifier or stretched while they are inside the cavity by using the group delay dispersion described in the previous section. In the first case, a combination of gratings stretches the pulse. In the second, optical elements inside the amplifier cavity stretch the pulse, mostly the transparent elements like the Pockels cell or the amplifier crystal. The stretched pulses have the same energy as the ultrashort pulses but much lower peak intensity since they have a longer duration. The amplification of stretched pulses is often referred to as chirped pulse amplification. After the amplified stretched pulses are dumped from the amplifier cavity, they need to be compressed again, which is done most often with a pair of gratings[4].

To avoid amplified pulses from running back into the seed laser and causing damage or instabilities, a Faraday isolator decouples the seed oscillator from the amplifier.

Figure 3. Setup of a regenerative amplifier. A Pockels cell combined with a quarter-wave plate and a thin-film polarizer (TFP) act as an optical switch. The Faraday rotator is used for separating input and output pulses. Courtesy of R. Paschotta (2008). Encyclopedia of Laser Physics and Technology, p. 632. Berlin: Wiley-VCH Verlag.

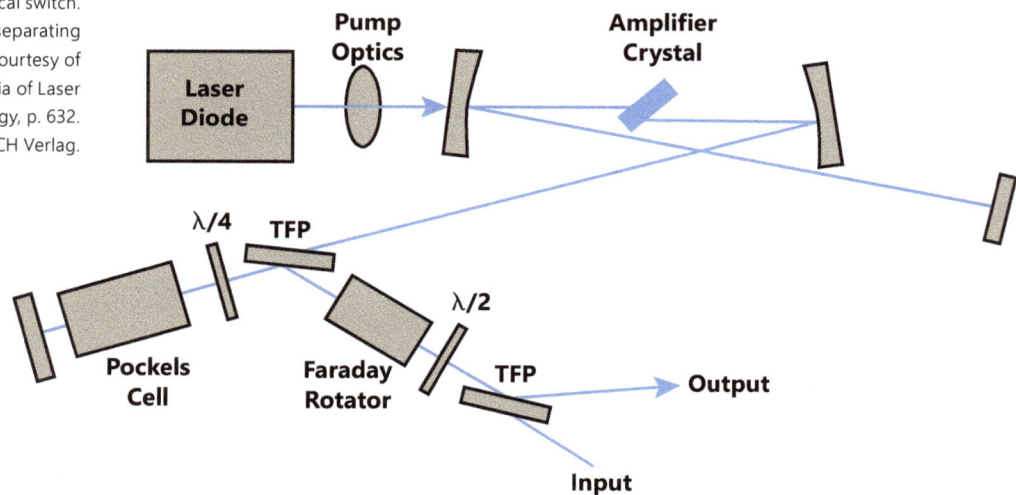

The key to fast switching

Particularly in laser material processing applications, ultrafast systems not only need high pulse peak powers but also they can benefit from high repetition rates that reduce the process time per work piece. The Pockels cell picks the pulses from a pulse train emitted by the seed laser and determines the repetition rate. Given a typical cavity length of 1.5 m and 100 round-trips, the pulse stays in the cavity for approximately one microsecond; theoretically, approximately 1 million pulses per second can be amplified. Historically, the piezo effect prevented these high switching rates. The Pockels cell consists of an electro-optic crystal that, by applying high voltage, rotates the polarization of one single pulse out of the 100-MHz pulse train. This single pulse is then coupled into the amplifier via a polarizer.

When high voltage is applied to the Pockels cell, the electro-optic crystal gets deformed and stressed due to the piezo effect. This stress will cause additional birefringence via the elasto-optic effect, which will disturb the originally desired electro-optic effect. Since the pulses emitted by the seed laser are only separated by approximately 10 ns, a very fast, precise switching Pockels cell is required, which means the piezo ringing should be as small as possible. So, electro-optic materials would be chosen in this case where the piezo effect is pronounced very weakly, like beta barium borate (BBO) or rubidium titanyle phosphate (RTP).

The switching voltage of BBO is much higher than for RTP; however, it has an easy-to-compensate group delay dispersion and a superb absorption coefficient, making it the perfect choice for high-power laser systems. Up to approximately 300 kHz BBO can be used relatively easily as a very precise switching electro-optic crystal. At higher frequencies, though, even small piezo ringing effects will cause trouble with the ringing of the crystal lattice dissipating as heat. For example, at 750 kHz the crystal can get heated up to 130 °C. This heat will cause thermal expansion and twisting of the Pockels cell housing, which results in misalignment of the crystal or damaged Pockels cell. Water cooling may drain the heat, but since water and high voltage do not mix well, the design of a water-cooled cell is too complicated to be economical.

New BBO Pockels cells have recently been developed with improved thermal management and very stiff and temperature-stable housing (Figure 2). Together with BBO crystals that have the smallest possible absorption, this Pockels cell represents a new benchmark in electro-optics for fast switching in the MHz range and the resulting high-average laser powers.

Figure 4. Qioptiq's low-outgassing Faraday isolator helps to increase the lifetime of encapsulated, high power and UV short pulse laser systems. Courtesy of Qioptiq, an Excelitas Technologies company.

The Faraday isolator

As mentioned before, the Faraday isolator is a key element in a MOPA system since it decouples the amplifier from the oscillator. In many regenerative amplifier configurations, a Faraday isolator is also used to couple the pulse into the amplifier and dump the pulse back out again (Figure 3).

Since laser powers have increased dramatically and UV conversion of the amplified short pulses has become more popular, laser manufacturers are placing greater importance on the low-outgassing properties of components within their systems. Typically, Faraday isolators are manufactured with many adhesives, which are among the materials with the worst outgassing behavior. Recently, adhesives have been avoided wherever possible and if they cannot be avoided, they have been replaced with vacuum-compatible adhesives, which resulted in a complete new series of low-outgassing Faraday isolators (Figure 4).

For years, femtosecond pulses have been the shortest events that could be produced artificially by humans and hence, the laser systems have been operated at the edge of technical and physical feasibility. Today, technical issues are steadily being resolved and prices are reducing per mW output power, opening up exciting new frontiers in femtosecond technology for a wide variety of new applications.

References

1. W. Koechner (1976). *Solid-State Laser Engineering.* Berlin: Springer Verlag, p. 87.
2. Ibid, p. 57.
3. R. Paschotta (2008). *Encyclopedia of Laser Physics and Technology.* Berlin: Wiley-VCH Verlag, p. 507.
4. C. Lupulescu (2004). Femtosekunden-Analyse und Rückkoppelungskontrolle von molekularen Prozessen in organometallischen und alkalischen Systemen, Freie Universität Berlin dissertation, p. 22.

Meet the author

Markus Fegelein earned his Diplom degree in physics from the Swiss Federal Institute of Technology in Zurich (ETHZ). During his 10-year tenure at Linos and later on Qioptiq, Fegelein has served as product manager for Faraday isolators, Pockels cells and electro-optic modulators.

Life in the Ultrafast Lane With Fiber Lasers

Ultrafast fiber lasers are proving less expensive and more comprehensive than traditional alternatives, expanding the range of applications. With cost-efficient, robust design, this group of lasers is advancing in ways that were not practical before.

BY HANK HOGAN, CONTRIBUTING EDITOR

For lasers, shorter pulses mean more precision and new applications. That's been true for decades. What's changed in the last few years is that prices for lasers with pulse widths a picosecond or less are dropping significantly. The price plunge has been pushed along by ultrafast fiber lasers, which are less expensive initially and cost less to operate than alternatives.

As a result, it's now possible to remove thin films, drill holes and otherwise process materials on a microscopic scale in ways that weren't practical before. These lasers can also slice and dice transparent materials like sapphire and glass. That further expands the range of possible applications and uses.

All of this adds up to growth of the ultrafast fiber laser market. Some of which comes at the expense of other laser technologies, according to René Kristiansen, product line manager for aeroPulse Ultrafast lasers at NKT Photonics A/S of Birkeroed, Denmark.

A laser-cut bicycle beside a 2 euro coin. Courtesy of Trumpf.

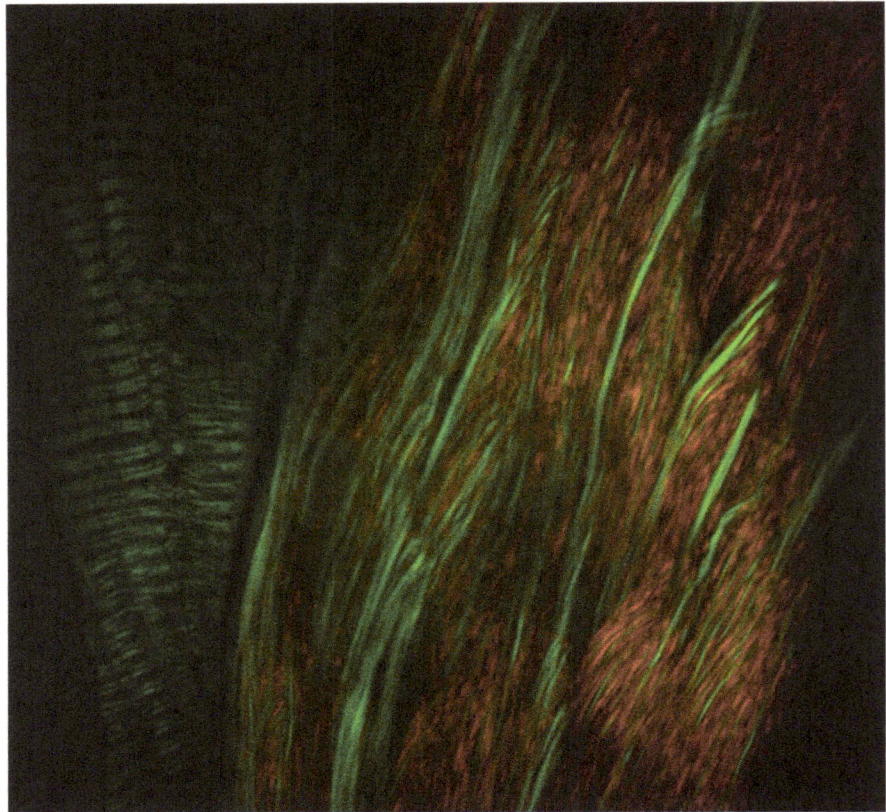

A two-color image of two-photon excitation done using an ultrafast fiber laser. Courtesy of Toptica.

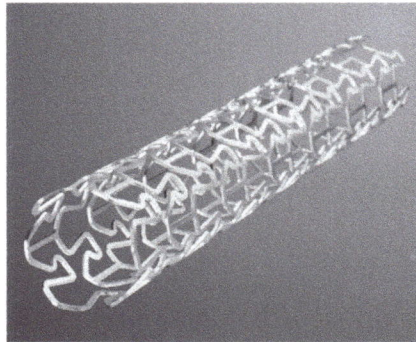

A laser-cut stent. Courtesy of Trumpf.

"In a lot of cases you're replacing UV longer pulse lasers, say nanosecond or picosecond lasers, with IR femtosecond lasers. In many cases, a shorter pulse IR laser can do the same job as a UV longer pulse laser used to do," Kristiansen said.

He noted that ultrafast fiber lasers have a cost-efficient design. One reason is that most major manufacturers have switched over to photonic crystal fiber within the last few years. This approach, according to Kristiansen, is both less expensive and yields inherent beam quality without requiring much tweaking or adjustment. That makes such lasers operationally robust.

For manufacturers, a benefit of this approach is that it's possible to take a single design and use it as a platform to quickly spin out multiple variants. In turn, that lowers costs for systems targeting specific applications.

As long as volumes continue to go up, prices for ultrafast fiber lasers will go down. However, future price drops may be somewhat mitigated. Manufacturers tend to increase power or add capabilities to laser designs over time, Kristiansen said. That adds to the cost and counteracts, to a degree, the price reductions generated by increased volume and moving further along the manufacturing maturity curve.

There are a number of ways in which capabilities can be improved. This includes power, either peak or average. There's also pulse width, wavelength, repetition rates, spectral bandwidth, beam shaping and other possible parameters. Often, adjusting one of these will impact the others and require making a trade-off — if the parameter can be adjusted without a great deal of work at all.

"Changing pulse width can be difficult," said Tim Gerke, North America sales and marketing manager at Fianium Inc., a subsidiary of Southampton, U.K.-based laser maker Fianium Ltd.

One exception is power. It can be upped by adding more pump diodes or by increasing the length of the fiber run. The first increases the starting point for the laser gain process while the second means there is more gain material and thus more gain, Gerke said.

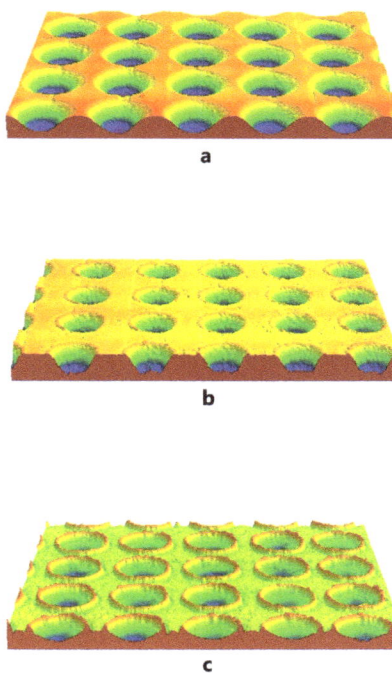

Surface structuring of bulk stainless steel sample using 4- (a), 46- (b), and 415- (c) picosecond infrared lasers. Each dimple was formed by 150 pulses, and the recast lip clearly increases for longer pulses within the picosecond range and demonstrates the increasing heat effects with increasing pulse width. Courtesy of Fianium.

Repetition rate, another parameter that might be adjusted without impacting others, runs up against outside constraints. Repetition rates could run as high as a few hundred kilohertz, making it theoretically possible to make materials processing faster and throughput higher.

But the beam must be directed to the right spot by a scanner. Many of these are designed with maximum rates of about a megahertz, more than fast enough for Ti:sapphire lasers, an older ultrafast laser technology. However, a megahertz rate of beam movement is too slow to take advantage of the highest repetition rate possible with ultrafast fiber lasers. Development work is underway on what are called polygon scanners, which should bump up the maximum usable repetition rate, Gerke said. He added that in other ways ultrafast fiber lasers offer some significant advantages over the incumbent technology. For instance, Ti:sapphire lasers operate in free-space cavities, get the pulses to the right place by bouncing them off mirrors, and require water cooling. In contrast, ultrafast fiber lasers produce a beam confined to an optical fiber and are air cooled.

Operating within a fiber means there is nothing to go out of alignment or require adjustment. As for air instead of water cooling, getting rid of water gets rid of one more complicating factor.

Both types of ultrafast lasers enable such applications as drilling holes in foil. The short pulse width makes this possible because the material is vaporized without redepositing, which makes the process cleaner. A key advantage of the ultrafast pulses is that they enable high peak power, which leads to pulses being absorbed by normally transparent materials, Gerke said. Consequently, a rapidly growing application and market is the processing of sapphire and glass found in smartphones.

The short pulse width of ultrafast lasers powers a whole array of applications that involve multiphoton absorption, according to Tim Paasch-Colberg, director of marketing at Toptica Photonics AG of Munich. The company makes ultrafast fiber lasers, with an area of concentration in biophotonics and the life sciences.

Multiphoton absorption is a nonlinear process in which, as the name implies, more than a single photon is absorbed at one time by an atom or a molecule. This is possible when a beam is concentrated into a small volume and its intensity is great enough. Hence, an ultrafast laser with a short pulse width and accompanying high peak power can make it happen. A fiber-based laser offers advantages in doing so because the systems are user-friendly and have a relatively low cost of ownership, Paasch-Colberg said.

The nonlinear nature of the process makes imaging deep within tissue possible because multiphoton absorption only happens in small volumes where the intensity is the highest. For this case, the power of the system can be minimal.

"A lot of applications require only moderate laser output of only a few tens of milliwatts, especially in nonlinear microscopy or industrial measurement," Paasch-Colberg said.

When materials processing (and not just imaging) is the object, more power is beneficial, he added. An example would be three-dimensional microlithography via two-photon polymerization, a process in which molecules form long chains. Ultrafast lasers enable this because the incoming pulses induce polymerization in a small volume. That spot can be moved as needed.

In these and other applications, reliability is an important feature, with scientists wanting an imaging system to work when desired and industrial users needing the same out of manufacturing systems. Fiber lasers offer a high level of reliability, in part because the technology was adapted from that used in highly dependable telecommunication, according to Paasch-Colberg.

"This results in high stability and maintenance-free operation, combined with a relatively low price in contrast to other ultrafast laser techniques," he said.

Ultrafast lasers also make the processing of smaller structures possible. This is partly because the heat-affected zone of such lasers is small, with the pulses being too brief for the heat they generate to diffuse very far into the material. The ma-

Internal scribe in sapphire wafer made using an infrared picosecond high pulse energy laser for singulation purposes. Courtesy of Fianium.

Glass slide marked internally using a picosecond high energy laser watermark. The mark is only visible under particular lighting conditions, and the degree of visibility is customizable. Courtesy of Fianium.

terial removed or deposited is likewise miniscule. This leads to greater precision in processing and the creation of smaller structures. A lower-cost ultrafast laser makes such an application more cost-effective.

Sometimes an ultrafast laser does something not possible otherwise, like manufacturing microstructures. But, in other instances the new technology is replacing traditional techniques. For instance, medical devices require marking for identification, such as by the use of a corrosion-free black mark. This is an example of ultrafast lasers replacing conventional techniques, said Max Kahmann, product manager at Ditzingen, Germany-based laser maker Trumpf. Looking toward the future, he noted that simply upping the repetition rate without adjusting other parts of the system, such as the scanner speed, may not yield much of a benefit. One reason is that without fast enough optics the pulses will simply overlap more as the interval between them goes down.

In another example where optics play a key role in performance, Kahmann noted that beam shaping increased the cleaving speed of glass in one application a hundredfold. So, future improvement in ultrafast laser system capabilities might be achieved by changes in components other than the guts of the laser system itself.

As for the continuation of the past trend toward cost reductions, the pace, which has been substantial over the last five or so years, could well accelerate, according to Kahmann, due to applications such as cutting sapphire and other transparent materials for smartphones, a rapidly growing use of ultrafast lasers. Applications like this represent large markets, and therefore offer opportunities to mature the technology. If so, the forecast trajectory outlines future cost trends.

"Especially when the lasers enter very large quantity markets the price reduction will be significantly large, maybe faster and larger compared to the past half-decade," Kahmann said. "After this large step — currently lots of markets are just doing this step — the price reduction will slow down, but on a much lower price level as we see it today."

New CO Laser Technology Offers Processing Benefits

The development of a reliable, high-power source of mid-IR laser light gives process developers an important tool with unique characteristics that will significantly impact a diverse range of applications.

BY ANDREW HELD, COHERENT INC.

The first carbon monoxide (CO) lasers were built more than 50 years ago. The technology showed promise for several reasons. First, CO lasers are inherently efficient in terms of their conversion of input electrical energy into light. For example, they are potentially about twice as efficient as the more commonly used carbon dioxide (CO_2) lasers. Also, CO lasers output in the 5- to 6-µm spectral range, whereas CO_2 lasers typically lase at 10.6 µm. This shorter wavelength provides processing benefits in many applications.

Given these advantages, why have CO lasers remained largely a laboratory curiosity until now, while CO_2 lasers, developed around the same time, have found widespread use in a diverse range of industrial processing applications? The answer is that CO laser technology could not be made practical, reliable and cost-effective enough for commercial use. In particular, early CO lasers required cooling in order to reach high output power (cryogenic cooling for very high powers). Also, while the first sealed CO_2 lasers could operate at high powers for hundreds of hours, early sealed CO laser lifetimes measured in just hours before output power dropped substantially.

This situation has now changed dramatically with the development of new CO laser technology that addresses these practical limitations. This has enabled the

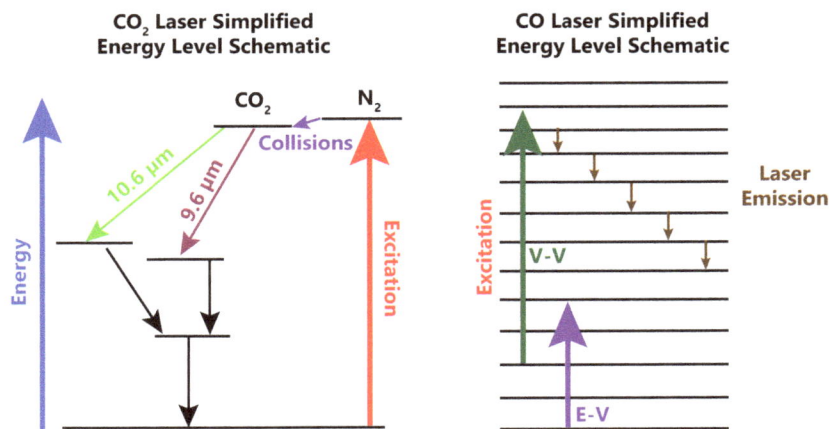

Figure 1. In a CO_2 laser, energy transfer from N_2 raises the CO_2 molecules into an excited state, which then decays into one of several possible lower states by emitting a photon. Nonradiative losses cause the molecule to return to the ground state. In a CO laser, several pumping mechanisms can raise the energy of the molecule to various excited states. It then cascades back down to the ground state, emitting several photons along the way.

production of sealed CO lasers that operate at very high output powers with excellent efficiency at room temperature and which demonstrate lifetimes in the thousands of hours range. This article reviews the technology behind the new generation of CO lasers and examines how their unique output characteristics lead to significant benefits in some important commercial applications.

CO laser development

In order to appreciate the inherent difficulties associated with developing and manufacturing a gas discharge laser like the CO laser, it is useful to examine the challenges met with the more familiar CO_2 lasers. The dynamics involved within fully sealed gas discharge lasers are complex, to say the least. Critical chemical

Figure 2. A CO laser with only 9 W of output power produced a clean, curved cut (6-mm radius circle) in thin glass (50-μm thick).

dynamic control involves the gas chemistry within and outside the laser discharge volume (including laser on and off), the chemistry between the gas and resonator materials (metals and ceramics), the chemistry between the many different materials used to construct the laser, and the chemistry involved with potential contaminants. Cleaning is critical within the sealed resonator tube.

Over the years, Coherent Inc. has learned how to control the chemical dynamics in fully sealed CO_2 lasers, enabling higher performance and improved reliability. Today's CO_2 lasers offer higher average powers and peak powers, as well as better power and pulse stability than ever before. At the same time, the reliability of these lasers has increased fivefold over the last 10 years. This can be attributed in large part to the ability to control the complex chemistry involved in CO_2 lasers: improved engineering design and development, as well as manufacturing process optimization, such as supplier development (coatings, electronics, etc.). The design improvements and critical manufacturing process optimizations for CO_2 lasers, developed over many years, are now being applied to the new CO lasers.

Figure 3. A 50-μm-thick glass substrate drilled with successively more pulses from a CO laser demonstrates the ability of this source to drill glass interposers, as well as drill micro dots on light guide panels used in display backlights. CO_2 drilling of this material typically results in heat-related cracking.

Gas dynamics in a CO_2 laser are a carefully controlled symphony of chemical activity. In CO_2 lasers, nitrogen gas is vibrationally excited via electron impact in a radio frequency (RF)-generated plasma discharge volume. The nitrogen then transfers its vibrational energy to the CO_2 molecules, which can now emit pho-

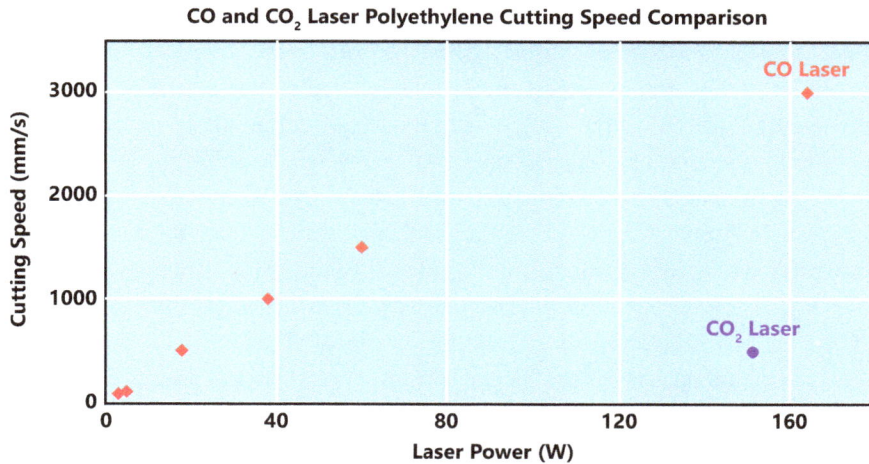

CO and CO_2 Laser Polyethylene Cutting Speed Comparison

Figure 4. Polyethylene cutting speed of CO lasers of various output power levels as compared with a CO_2 laser. A 20-W CO laser cuts at about the same speed as a 150-W CO_2 laser.

tons (9.6 or 10.6 µm) as they drop to lower energy vibrational states. From these lower vibrational energy states, the CO_2 molecules transition ultimately to the lowest energy state through collisions with helium, that is, through vibrational-translational (V-T) energy transfer. From this lowest energy state, they can then be re-excited.

In addition to the "lasing" gas dynamics, there is also the chemical decomposition and recombination of CO_2, which is carefully controlled by trace elements in the gas mix. Pressure and resonator geometry optimization also plays a critical role in controlling the gas dynamic and chemistry. Further, contaminants missed through the cleaning process or created in the chemical pool can disrupt the gas dynamic chemical balance and adversely affect the lasers' performance.

The chemical dynamics of the CO laser are different but equally complex. Vibrational excitation of CO molecules takes place directly via electron impact from the same RF sources used to create the discharge plasma in a CO_2 laser. However, CO lasers don't necessarily need nitrogen. Once vibrationally excited, CO molecules can further excite other such molecules through vibrational-vibrational (V-V) transfer.

Laser emission comes from successive transitions between pairs of populated vibrational levels in the diatomic energy-level ladder. Each transition to a lower energy level then becomes the upper state for a subsequent transition — this is the definition of a cascade laser. As with the CO_2 laser, there are many different processes competing with effective lasing that need to be accounted for. Where the helium V-T cooling is critical for CO_2 lasers, in CO lasers this can actually compete with the efficiency of the V-V CO pumping process. This is partly why early CO lasers were often cryogenically cooled. Room temperature operation was not practical, until now.

Mid-IR wavelength advantages

Much of the interest in CO lasers derives from the fact that its midwavelength-IR output offers two important advantages, in terms of applications, as compared to the long-wavelength IR output of the CO_2 laser. The first of these is the fact that many materials exhibit significantly different absorption at the shorter wavelength, leading to disparities in their processing characteristics that can be exploited. Specifically, in cases where the absorption is higher at the shorter wavelength, material can be processed more efficiently using lower laser power and with a smaller heat affected zone (HAZ). This stronger absorption occurs in many metals, films, polymers, PCB dielectrics, ceramics and composites. On the other hand, when the transmission is higher at the shorter wavelength, the light penetrates farther into the material, which is also sometimes advantageous.

The other major difference is that shorter wavelengths can be focused to smaller spot sizes than longer wavelengths due to diffraction, which scales linearly with wavelength. The final spot size depends on working distance and the numerical aperture of the focusing lens. The theoretical, diffraction-limited spot size for 10.6-μm CO_2 lasers is about 55 μm, while the minimum spot size achieved in practice in industrial applications is 80 to 90 μm. The 5-μm CO laser can reach theoretical spot sizes of about 25 μm under the similar focusing conditions, with practical spot sizes in the 30 to 40 μm range. As a result, the CO laser spot can have a power density (fluence) that is four times higher than the CO_2 laser. The higher power density, when combined with stronger absorption in some materials at 5 μm, enables these materials to be processed with a CO laser at significantly lower powers.

Diffraction also dictates that a shorter wavelength spreads more slowly over distance, leading to improved depth of focus. Specifically, depth of field — the distance over which the beam size increases by a factor of $\sqrt{2}$ — is related to laser wavelength (λ) and input beam diameter (D) by:

$$Depth\ of\ field = \frac{\pi D^2}{2\lambda}$$

Thus, for a given input beam diameter, depth of focus increases as wavelength decreases. The benefits of a longer depth of focus include higher aspect ratio processing and an increased process window, especially for materials with uneven surfaces and/or variations in thickness.

Glass cutting

CO_2 lasers are already employed in cutting thin ($<$ 1-mm-thick) glass sheets, namely the specialty glass (strengthened and nonstrengthened) used in many smartphone and tablet displays. The 10.6-μm output of the CO_2 laser is strongly absorbed by glass, whereas absorption around 5 μm is much lower. But this actually leads to significant potential advantages in this application.

In CO_2 laser-based glass cutting, the light is absorbed strongly at the surface, generating heat that must then diffuse into the bulk material. After laser exposure, a jet of water or air is used to thermally shock the material and create a precisely controlled scribe or vent. Mechanical means are then employed to actually break the glass along the laser-scribed line.

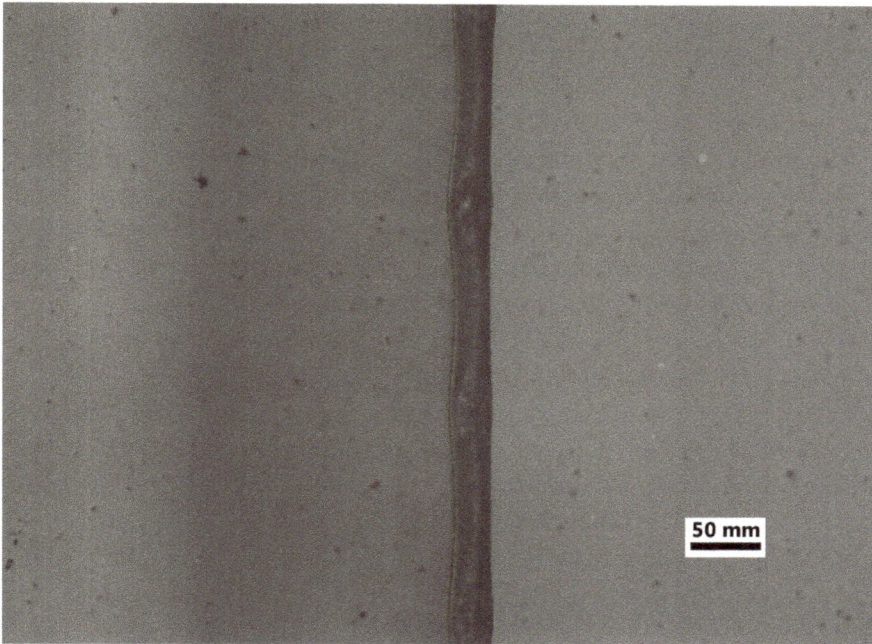

Figure 5. Top view of the edge of a 20-µm-thick polyethylene sheet, cut using a 164-W CO laser at a feedrate of 3000 mm/s. The laser produced a clean cut edge, with a heat affected zone of about 29 µm.

The overall process is much the same with the CO laser; however, the lower absorption allows the light to penetrate much further into the bulk material. Heat is introduced to the glass directly and does not rely on diffusion from surface. Tests have been performed indicating that this eliminates surface melting, avoids the creation of cracks and produces no residual stress in the glass. The results are a better quality scribe yielding a stronger cut piece, together with a wider process window, for the manufacturer.

The other exciting aspect of CO lasers in glass cutting is their ability to support the cutting of curves. This is of particular significance in smartphone display applications, as curved or shaped corners are often required to accommodate buttons, controls, LEDs and camera lenses. CO_2 lasers are typically limited to cutting glass in straight lines because their round output beam must be reshaped into a long, thin line in order to distribute the intense heat generated at the surface. In contrast, the lower absorption of the CO laser allows its round beam to be used directly without adverse heat effects.

The CO laser also enables processing of very thin glass (below 300-µm thick). This material is almost impossible to cut mechanically and is also challenging to process using the CO_2 laser. In this case, the CO laser can cut completely through the glass, eliminating the need for a subsequent mechanical breaking step, which is particularly difficult to accomplish with very thin glass.

Drilling glass holes

Another important glass processing application is drilling for glass interposers used in 2.5D and 3D advanced circuit packaging techniques. This application takes advantage of both the superior focusability and lower absorption of 5-µm light in glass. Specifically, it enables very small holes to be drilled in glass with precise depth control and no heat damage or cracking.

Film cutting

Polyethylene (PE) has strong absorption at 3.5 µm, with mild transitions at 7 µm and ~14 µm. Unfortunately, no high-power lasers are available at any of these wavelengths, so laser cutting of PE has not been practical in the past.

While low-level absorption can always be driven with high powers at CO_2 laser wavelengths, the residual heat generated by the unabsorbed light creates unacceptable collateral damage. But the higher focused fluences achievable with the CO laser avoid this problem, enabling PE to be effectively cut.

Testing has shown that thin (20 µm) PE can be cut at speeds beyond 3000 mm/s with a limited HAZ (30 µm). CO_2 laser cutting of this material at similar power levels reaches 500 mm/s at best and creates 500 µm of HAZ. At 60-µm thickness, the CO_2 laser simply cannot cut the PE without destroying the material, whereas the CO laser cuts with acceptable speed and cut quality.

Meet the author

Andrew Held holds a Ph.D. in physical chemistry from the University of Pittsburgh. He is director of marketing at Coherent for the company's CO_2 laser products.

Ultrafast Lasers Enable Improvements in Microprocessing

Truly athermal processing is one of the advantages of ultrafast laser machining over longer-pulse lasers.

BY TIM GERKE, FIANIUM INC.

Ultrafast lasers provide an avenue into a number of new microprocessing applications because of their capability for athermal ablation and their extremely high peak power. Many materials and applications can be susceptible to adverse heating effects such as melting, peeling, chipping, melt splatter and substrate damage when processed with CW and long-pulse lasers. With picosecond and femtosecond lasers, however, these undesirable side effects can be mitigated or eliminated completely. In addition to decreasing heating effects, using ultrashort pulse widths can increase process efficiency and even enable new applications — such as internal marking or scribing of bulk transparent materials — that otherwise are impossible with CW or longer-pulse-width lasers.

These benefits can be demonstrated using an array of high-energy fiber lasers with pulse widths of 2 to 400 ps and pulse energy up to 125 µJ. The lasers used in the demonstrations discussed here are installed in an applications lab in Portland, Ore., on 2D galvanometer scanner-based processing systems, as well as in a fixed-focal-spot station used for creating very small laser spot sizes. Both processing systems are computer controlled and synchronized with the laser system for full control and arbitrary pattern processing.

This article discusses, by way of particular representative examples, the advantages of ultrafast lasers over longer-pulse lasers for microprocessing applications. Actual commercial use may require the licensing of intellectual property, depending on details of the application, the system assembled or the particular method practiced.

Athermal, high-quality processing

Many microprocessing applications can be executed with nanosecond lasers — but with a significant loss in the quality of the result. Such lasers fundamentally impart much more heat to the material and result in melting, cracking, surface composition changes and other detrimental side effects. With ultrafast laser machining, however, truly athermal processing can be achieved on any material and in any application. One particularly good match for high-energy picosecond lasers is thin-film processing. In this application space, lasers with very moderate pulse energy have been shown to be capable of high-speed material removal with excellent qualitative results, while nanosecond lasers often struggle to achieve equivalent quality and are prone to a number of defects. The heating of thin-film parts due to the long nanosecond pulses can cause an array of detrimental side effects from recast lips, melting, chipping or substrate cracking or damage.

Figure 1 demonstrates one such defect that can occur in thin-film microprocessing with nanosecond lasers. In the figure, a thin film of molybdenum on a thick glass substrate was laser-scribed with singular pulses in a back-side (superstrate) geometry, where pulses are focused through the glass substrate onto the back of the film. The film is then removed in a liftoff process. In the 10-ns case (Figure 1c), small bits of material are peeling off around the edges and protruding vertically from the film surface, and one such example can be seen on the left edge of the scribe just below center. For conductive materials, this sort of defect is often catastrophic for device functionality.

For the 100-ns case (Figure 1d), a lot of material has been deposited onto the

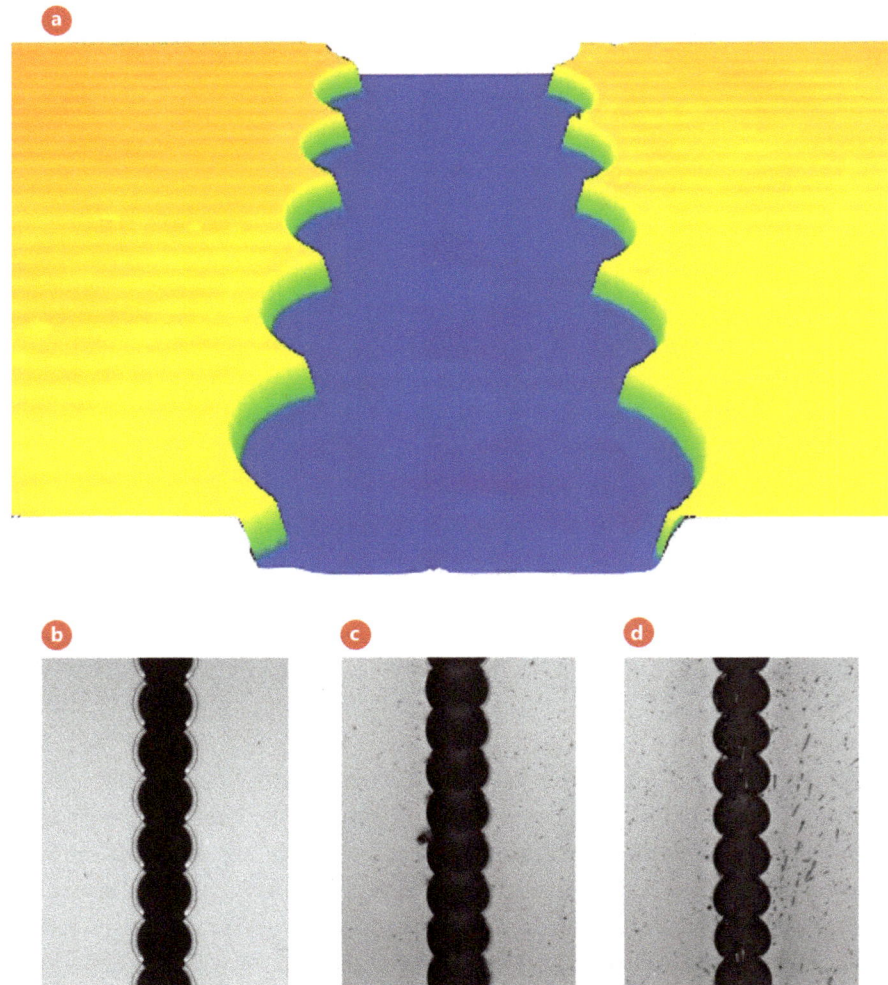

Figure 1. Scribes of thin metal film on glass using (a) 415-ps, (b) 46-ps, (c) 10-ns and (d) 100-ns pulses. Images courtesy of Fianium.

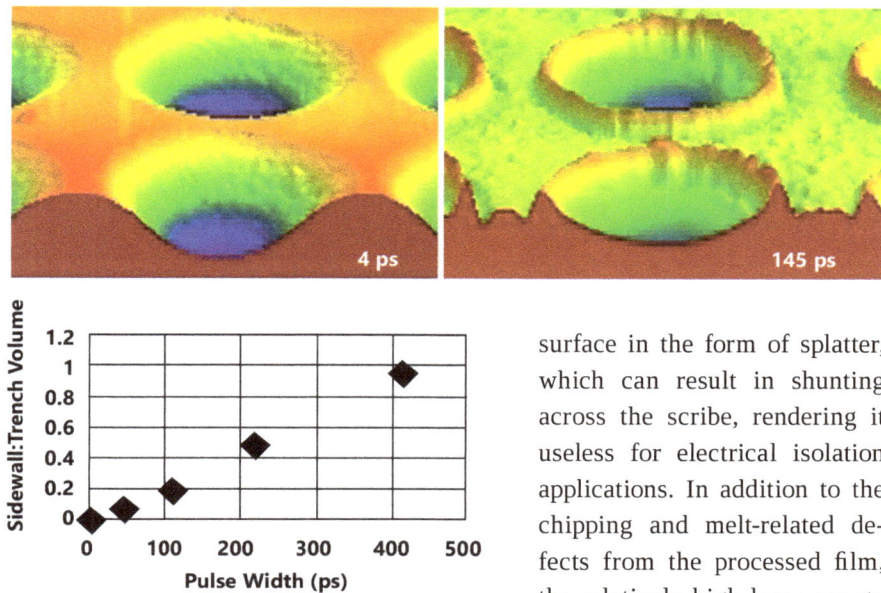

Figure 2. 3D surface profile images of bulk steel surface textured by multiple pulses per dimple using 4- and 415-ps laser systems (top). A plot of the sidewall:trench volume ratio as a function of pulse width (bottom).

surface in the form of splatter, which can result in shunting across the scribe, rendering it useless for electrical isolation applications. In addition to the chipping and melt-related defects from the processed film, the relatively high laser energy resulted also in microcracking of the underlying glass substrate, which can be an impediment to applications where the structural integrity of the substrate is of concern. Overall, there are significant issues with the results achieved from processing this thin metal film with nanosecond systems.

On the same material, we also investigated the results of picosecond pulse widths from 4 to 415 ps using Fianium laser systems, and the picosecond results were qualitatively far superior to the nanosecond results. No modification to the substrate was observed, and no film chipping, cracking or melting was apparent. Processing debris was also not observed. Figures 1a and 1b show the results of 415- and 46-ps pulse widths, respectively. Both results demonstrate typical high-quality, fully isolating scribes of the thin-film material using single laser pulses. Such results are achievable even up to well over 1 m/s with relatively low power (<2 W) lasers or over 10 m/s with higher-power lasers such as Fianium's Hylase system (8-25 W).

Improvements from ultrafast laser solutions are not limited to thin-film applications, however. Applications involving microprocessing bulk materials, such as metals and semiconductors, can also significantly benefit from ultrafast laser sources. For short (<20 ps) pulses, many of these materials can be machined athermally with no melting. With longer-pulse lasers, even still within the picosecond regime, the thermal effects can become apparent. Figure 2 demonstrates that, even with subnanosecond pulses, melting occurs and results in a recast lip on a laser-surface-textured bulk steel part. When the identical process is run with shorter 4-ps pulses, however, no recast is observed, and more material is removed to form deeper dimples.

The plot at the bottom of Figure 2 demonstrates the trend of the volume of material in the recast sidewall, relative to the removed volume. At 4 ps, where there is no sidewall, the ratio is zero. As the pulse width increases to 415 ps, the ratio

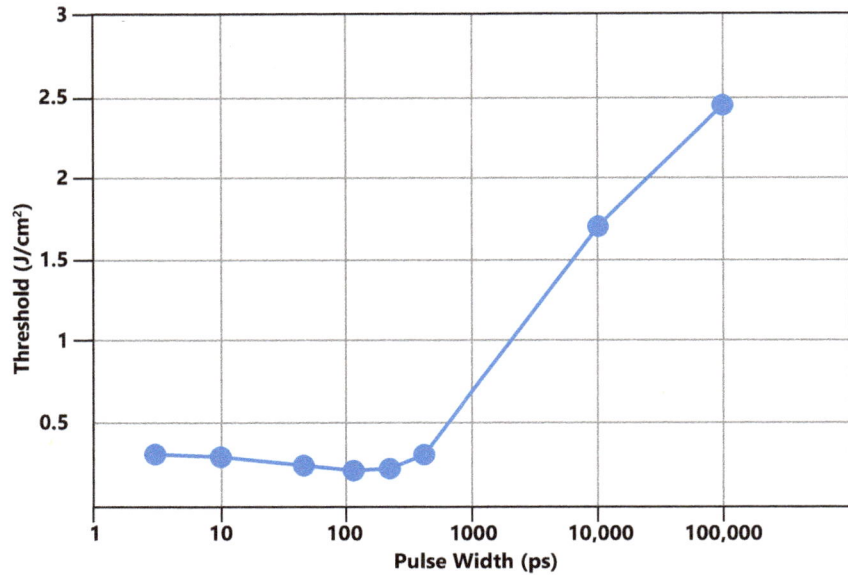

Figure 3. Plot of threshold for thin-film removal vs. pulse width. The removal threshold is flat for the picosecond regime and increases dramatically into the nanosecond regime.

increases to nearly unity, where the volume of material removed is approximately equal to the sidewall volume. At this value, the material in the trench is not actually being removed from the substrate; instead, it is simply being pushed to the side.

Increase in efficiency

In addition to avoiding heat-related defects, ultrashort pulse widths can improve processing results through increased efficiency. The long interaction time associated with nanosecond- and longer-pulse lasers results in more energy being wasted, because heat results in melting and can also quickly diffuse out of the modification area. Ultrashort pulses, on the other hand, have such short interaction times that virtually all of the energy applied by the laser pulses is directly utilized in the ablation mechanism and results in a more efficient process.

One example of this trend: The laser energy required to scribe or pattern thin films on bulk substrates increases with pulse width. This type of application is of-

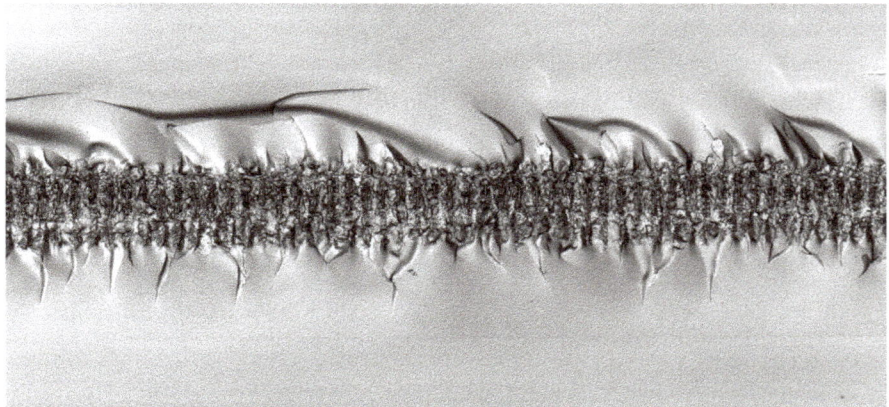

Figure 4. Break edge of a sapphire wafer that has been scribed internally.

30

ten utilized in a liftoff or superstrate processing geometry, where the laser energy is absorbed in the film-substrate interfacial region. Some small fraction of the film is vaporized, and the resulting vapor pressure trapped between substrate and film ejects the film within the processing area.

The plot of experimentally measured thin-film removal threshold versus pulse width (Figure 3) demonstrates that, in the picosecond regime, the removal threshold of thin molybdenum films is around 0.25 J/cm². When the pulse length increases into the nanosecond regime, the removal threshold increases significantly, requiring an order of magnitude more pulse energy than the picosecond case for the same removal area.

This trend demonstrates that a 10- to 100-ns laser would need to have approximately 10 times more pulse energy and average power to achieve equivalent process throughput relative to a picosecond laser system. Based on Figure 1, the nanosecond results would also be qualitatively inferior to the ultrafast laser results. This trend of improved efficiency with shorter pulse width is not specific to metal or other opaque thin films. Similar trends were also observed for a variety of thin transparent conducting oxide films, which are very popular thin-film materials. It has also been previously reported in academic articles demonstrating improved bulk material removal with shorter pulse width.

New capabilities from ultrafast lasers

Ultrashort pulse widths are also important for applications involving nonlinear absorption, where peak power is instrumental. In these types of applications, the high peak power achievable by ultrashort-pulse lasers can initiate a nonlinear absorption, either to ablate a material that typically would not respond to the laser wavelength, or to penetrate a transparent material and cause modification internally.

One such application that has experienced recent popularity is internal scribing of sapphire wafers and glass. This application requires focusing a laser that is transparent to the material inside the bulk and providing sufficient peak-power density to ablate the material via nonlinear absorption. The ablation process provides sufficient energy to create small microcracks in the material, and stacking many pulses side by side can create a line of weakness (Figure 4) to guide a subsequent break. Alternatively, with sufficient laser energy and power, the process can be optimized such that the cracks span the entire depth of the material and create a full cut of the material. Both cutting and scribe-and-break laser processing provide orders of magnitude increase in process throughput and improved reliability/ yield over mechanical methods such as diamond saws and mechanical scribes.

Another application that benefits from the high peak power and short pulse width of ultrafast lasers is internal marking. As with sapphire scribing, if short

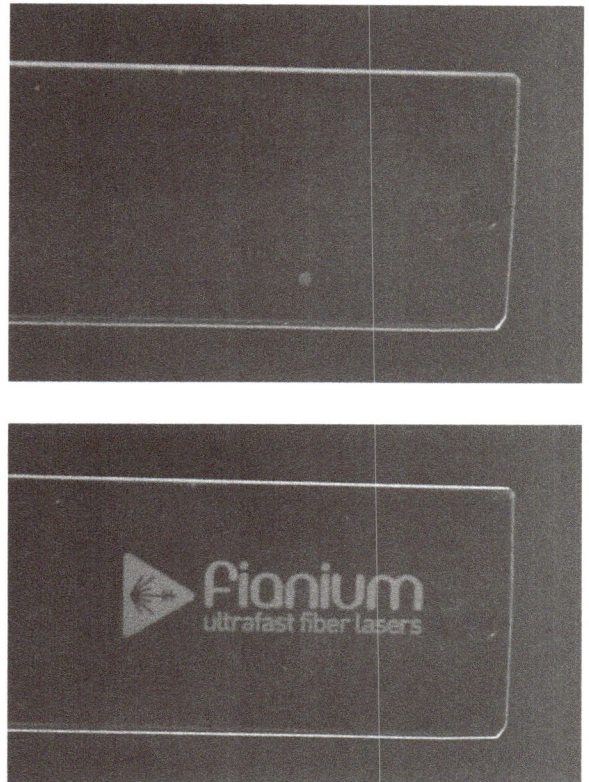

Figure 5. A sample marked with an easily visible watermark. Under one viewing angle (top), the mark is not visible, but at a second angle (bottom), it can be seen clearly.

pulses are focused inside a glass substrate (or other transparent material) and the energy level and pulse-to-pulse overlap are appropriately tailored, semivisible watermarks can be created. These marks are visible only under specific illumination conditions; they can be made easily visible or virtually impossible to detect. Figure 5 demonstrates one sample marked in this manner under two illumination conditions: one in which the image is not visible (top) and the other in which it is clearly visible (bottom). The only aspect that changed between the images is that the sample was rotated by 15°. This type of marking can be conducted only with short-pulse lasers. When attempted with longer pulses, even still well within the picosecond regime, the marks become microscale cracks that are visible under any viewing angle.

Meet the author

Tim Gerke is laser applications engineer at Fianium Inc. in Lake Oswego, Ore.

Better Lasers, Better Machining

Greater uptime, faster cutting and increasing affordability make fiber lasers the first choice where CO_2 lasers once reigned supreme.

BY HANK HOGAN, CONTRIBUTING EDITOR

When it's time to cut, weld, ablate, mark or otherwise machine materials, manufacturers are increasingly turning to lasers. Falling system costs and better resulting product quality are two reasons why. Still, the need exists to further lower the cost of laser machining while expanding the range of materials that can be handled. For that, there's progress in fiber and other laser technologies from greater power, more precise processing and new wavelengths.

An illustration of these trends can be found at Mazak Optonics Corp. The company, which has its North American headquarters in Elgin, Ill., does not make laser engines but does incorporate those supplied by well-known vendors to make laser-cutting equipment used at large companies and independent job shops.

Mazak Optonics has seen a significant change in laser machining technology over the past few years, according to Marc Lobit, general manager of sales support operations. "Laser-cutting technology is still shifting to fiber [lasers], with about 75 percent of our machines sold in 2015 being fiber," he said.

Lobit noted that five or so years ago a majority of the company's products used CO_2 lasers. Fiber's higher reliability and uptime, along with its associated decrease in maintenance costs as compared to CO_2 lasers, have driven the change.

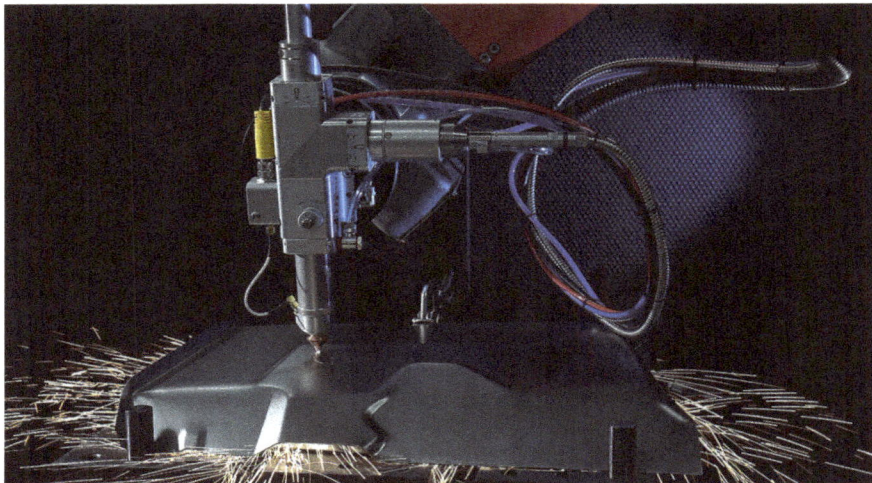

Fiber lasers, like the one shown here cutting an automotive part, are increasingly used for cutting and other machining. Courtesy of Coherent.

A big reason for this is the beam delivery system, which is straightforward for fiber lasers. In contrast, CO_2 lasers employ mirrors to get the beam to the business end of the cutting machine, which leads to relatively high upkeep.

"[CO_2 lasers] were very labor-intensive and expensive to maintain." Lobit said. "Mirrors would become damaged. They would wear out over time from the heat that's reflecting off of [them]."

The transition to fiber laser technology brought about other advantages, he added. The ytterbium fiber lasers output a beam at 1070-nm wavelength, about a 10th of that of a CO_2 laser. The result of this difference — the greater uptime and other differences — was faster cutting using a fiber laser, and better results, particularly when it came to edge quality. Consequently, a single fiber laser cutting machine can replace two to three of the older systems.

Laser machining does not completely replace and often is complementary to the traditional, mechanical approach, Lobit noted. Mazak Optonics is part of Japan's Yamazaki Mazak Corp., a manufacturer of laser and mechanical machine tools and systems. The company's customers may use lasers to cut but may go the traditional route for machining shapes or adding fine detailing.

As for the future of laser machining, more power is on its way. Today, Mazak Optonics has products ranging up to 6 kW. "Higher power is starting to hit the market. This enables thicker cutting, but probably the most important benefit will be faster cutting for 1/4- to 1/2-in. thicknesses," Lobit said.

Laser manufacturers are also coming out with new wavelengths. This is important because not all materials react the same way to a given wavelength, pointed out Tracey Ryba, product manager for laser systems and OEM lasers at Trumpf Inc. The laser manufacturer has its U.S. headquarters in Farmington, Conn.

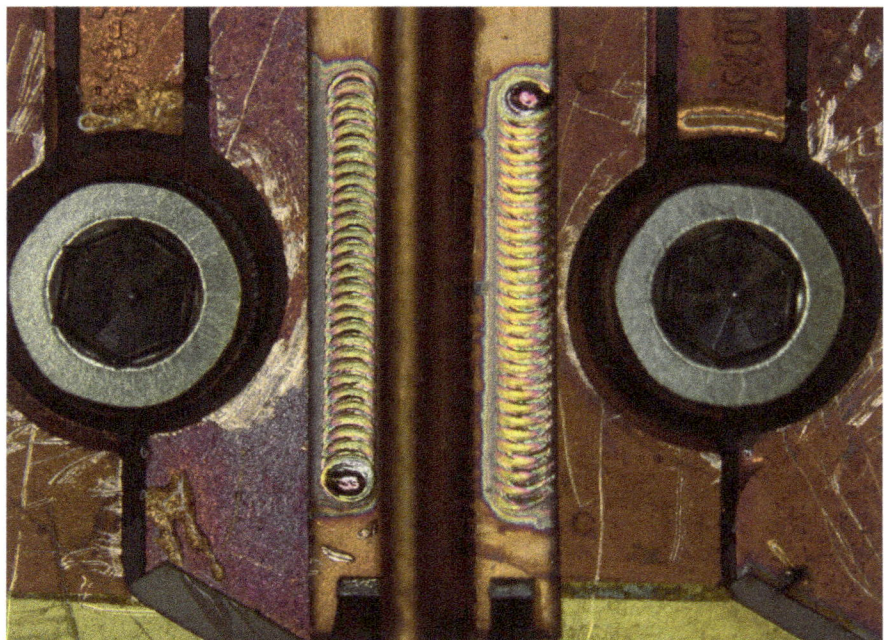

By doubling frequency to the green portion of the spectrum, lasers can weld copper, even thin layers, without damage to ceramic material lying below. Courtesy of Trumpf.

Global Fiber Laser Market 2014-2019 ($ Millions)

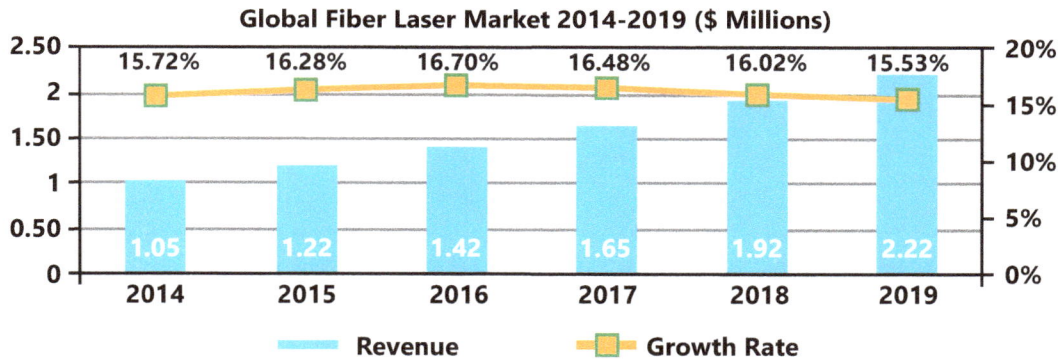

	2014	2015	2016	2017	2018	2019
Growth Rate	15.72%	16.28%	16.70%	16.48%	16.02%	15.53%
Revenue	1.05	1.22	1.42	1.65	1.92	2.22

Revenue   Growth Rate

For instance, CO_2 lasers are good for producing perforated holes for easy separation of food packaging plastic bags. As for metals, a wavelength around 1000 nm generally works well. However, copper is highly reflective at that wavelength. So, Trumpf developed a frequency doubled, pulsed green disk laser for welding copper, Ryba said.

High precision = less processing

As for the future of laser machining, it's instructive to look at the past. According to Ryba, fiber and disk lasers have largely replaced CO_2 lasers because they are more reliable and easier to maintain.

Both fiber and disk lasers depend upon a diode to pump the lasing material. Today, the beam quality or power (or both) of a diode laser makes it unsuitable for many machining applications, but Trumpf expects continued progress will change that. The push will be to simplify the laser system and go the direct diode laser approach, Ryba said.

This is because when it comes to laser versus traditional machining, ongoing cost cutting changes the balance and favors the laser approach, he added. Over the past seven or eight years laser systems have come down about threefold in cost per delivered watt. That drop in price cannot continue — unless the system can be further simplified. Going to a diode laser instead of a diode-pumped laser could do that.

It also could be that even without additional cost cutting that laser machining is more cost-effective than other approaches, due to the elimination of manufacturing steps.

"Part of the advantage of laser welding is you don't have to do so much post processing typically," Ryba said. "Compared to traditional machining, the laser is high precision."

For another example illustrating trends and developments in laser machining, consider Boeing. The Chicago-based aerospace giant has been laser cutting materials for over a quarter-century, said William Schell, Boeing associate technical fellow. Today, the company

The global fiber laser market was valued at $1.05 billion in 2014 and is expected to reach $2.22 billion in 2019, growing at a compound annual growth rate of 16.20% during the forecast period, according to Technavio. Courtesy of Technavio.

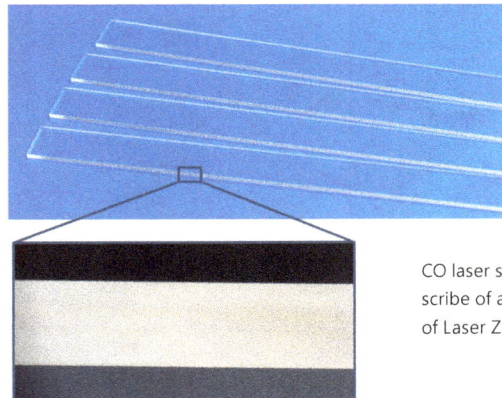

CO laser separation after laser surface scribe of a borosilicate glass. Courtesy of Laser Zentrum Hannover eV.

uses laser machining to fabricate the ducting system for its airplanes, replacing conventional milling, band saw and shear/nibbling equipment.

"The lasers offer exceptional speed and accuracy over other cutting processes, particularly on nickel-cobalt alloy materials, which are difficult to cut," Schell said.

He added that the company's technology research organization designed, implemented and patented a laser trim machine for trimming duct details. With it, a mechanic scribes a mark, puts it in the field of view of a camera and then a laser rotates around the stationary part, cutting off the excess to the scribe mark.

Boeing's laser trim machine addresses process limitations with conventional cutting methods and the safety issues brought about by trying to manually handle duct trimming. Schell noted that the use of a laser was particularly beneficial when an increase in the thickness of the duct walls was needed to meet design requirements. The thicker walls began to damage the conventional cutting equipment.

Asked about issues associated with laser machining, Schell mentioned one that impacts all manufacturing operations. "While lasers offer unique advantages over traditional methods, they also come with their own set of safety challenges. Being able to cost-effectively and safely implement the use of lasers in a factory environment — a production work cell — can be challenging," he said.

When looking at the future of laser machining, Frank Gaebler, director of marketing at Coherent Inc. of Santa Clara, Calif., points to several trends. One is the emergence of increasingly powerful fiber lasers. Due to their nature they are

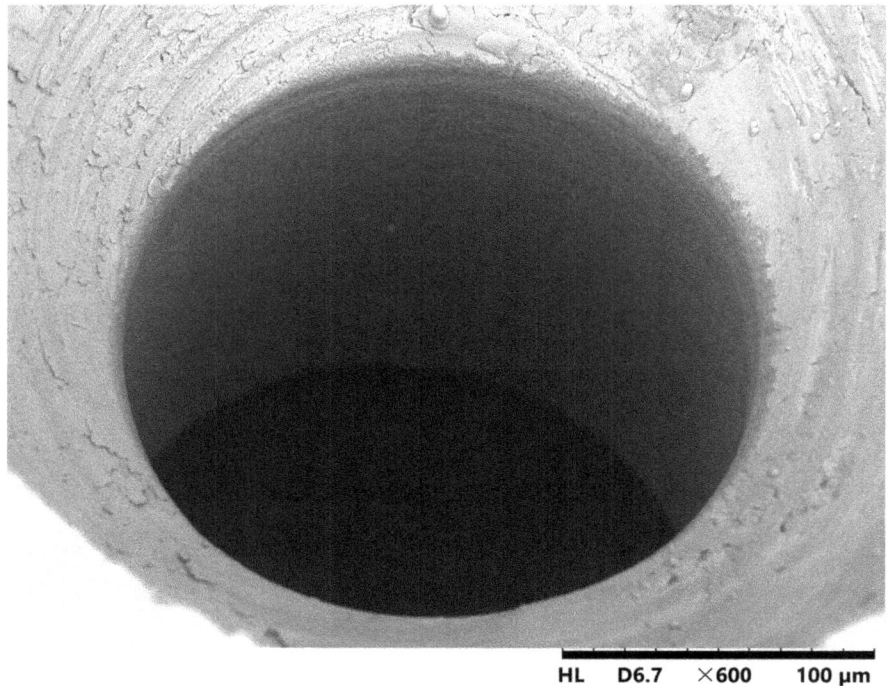

A gasoline injector nozzle drilled by a Coherent Monaco Laser through 250-µm-thick stainless steel. Courtesy of Coherent.

HL   D6.7   ×600   100 µm

inherently more reliable and cost less to operate than the old laser machining workhorse, the CO_2 laser.

However, the old standby is still best when working with very thick materials. "The shipyard business and these places where you have to cut plates instead of sheets, then the CO_2 laser might still be the best choice," Gaebler said.

Another important development is the advent of affordable ultrafast lasers. With pulse widths in the femto- and picoseconds, such lasers enable machining on a microscopic scale, something that can't be easily done, if at all, using traditional methods. Because the pulse widths are so short, the heat-affected zone in the material is negligible, and that gives laser processing an advantage.

For instance, ultrafast lasers are used to precisely drill holes of differing shapes and sizes in fuel injector nozzles. This buys better combustion of fuel, leading to greater efficiency, more power or some combination of both.

Another example are stents and other medical devices that go into the body. These are made of biocompatible polymers and other materials that have to be finely machined, presenting challenges to manufacturing methods not based on ultrafast lasers.

A third trend is the development of new wavelengths. Coherent, for instance, has come up with a CO laser that produces a 5-µm wavelength output. Because it is half the wavelength of a 10-µm CO_2 laser beam, it can be focused to half the spot size, which ups the intensity fourfold.

As to why a new wavelength might be needed, Gaebler noted that some materials do not work well with current lasers. Pigment-free polyethylene films, for example, have in the past not been processed well with lasers due to limited material absorption, too low a beam intensity or both.

"But at 5 µm it works pretty well. They can cut it with high reliability and very little heat-affected zone," Gaebler said.

Laser Welding Joins the Lightweighting Trend

Automotive manufacturers turn to mixed materials, lasers and robots to reduce the weight of vehicles.

BY JAMES SCHLETT, EDITOR

To return the Cadillac brand to the global prestige luxury stage, General Motors threw as much as it could — 13 materials in all — into the sedan's body to make it the lightest vehicle in its class. And then to join this "crazy quilt of aluminum, steel, magnesium and plastic," as one industry watcher called it[1], GM employed the auto industry's most comprehensive and advanced mixed-materials manufacturing techniques, which are deployed by an army of 28 robots that weld the inner and outer frames of the vehicle (Figure 1).

Primarily holding together the body of this sleek, lightweight car — as well as possibly the U.S. automotive industry's future joining strategy — are flow drill screws, self-piercing rivets, first-of-their-kind aluminum resistance spot welds and aluminum laser welds (Figure 2). Through this mixed-material approach, GM managed to reduce the Cadillac CT6's weight by 198 pounds, as compared to what it would have weighed with an all-steel body.

Figure 1. The aluminum roof of General Motors Co.'s 2016 Cadillac CT6 is welded with a laser to the sedan's bodyside joints. Courtesy of General Motors.

The CT6, which began production at GM's Detroit-Hamtramck plant in January 2015, is so far this year's highest-profile vehicle to jump on the so-called light-weighting trend. Lightweighting relies heavily on replacing steel with aluminum and is largely driven by federal regulations that call for stricter crash standards and for vehicles to achieve 54.5 miles per gallon by 2025.

Last year, the industry's lightweighting star was Ford Motor Co.'s F-150 pick-up, which shed 700 pounds largely by replacing steel for its body with aluminum (Figure 3). The aluminum roofs of both the F-150 and CT6 are welded to bodyside joints with lasers. GM also used lasers to weld the sedan's doors and its three-piece decklid joints, which resulted in a more curved trunk than a two-piece part, according to Elaine Garcia, GM's engineering group manager for advanced technology and welding.

"We laser-welded the aluminum roof and decklid because it enabled a certain appearance in the product design, and laser-welded the aluminum doors because it allowed for shorter weld flanges and a larger door window opening," said Garcia.

Ford's all-aluminum body approach for the F-150 is considerably different from GM's mixed-materials strategy (Figure 4). And just as there is much debate over which approach will be adopted by competitors, there is also much deliberation over what joining technologies automakers will favor most for the manufacturing of their lighter, sturdier vehicles. This is the time for industrial lasers to cut through the competition or get left behind by alternatives such as adhesives and spot welding.

Aluminum on wheels

Aluminum's weight is roughly a third of steel's, though they are both similar in terms of strength. The Aluminum Association's Aluminum Transportation Group (ATG) estimates that for every 10 percent of weight reduction attributable to the

> "With its high processing speeds, low heat input and resultant low distortion, and overall flexibility of application, this process [laser welding] has become a key assembly technology in the automotive industry during the last two decades. Because laser welding creates excellent quality joints, it is typically used in applications where weld appearance is critical."
>
> — *Doug Richman, technical committee chairman of the Aluminum Association's Aluminum Transportation Group*

Figure 2. GM reduced the Cadillac CT6's weight by 198 pounds by using a variety of materials, such as steel aluminum and magnesium, and joining them with several technologies, including proprietary aluminum resistance spot welding and laser welding. Courtesy of General Motors.

Aluminum Laser Welding Options

"Replacing steel with aluminum significantly changes the entire production process," said Tim Hurley, the global automotive segment manager for The Lincoln Electric Co. in Euclid, Ohio.

For the Ford Motor Co., that is an understatement. In 2014, the company closed its Dearborn, Mich., assembly plant and spent eight weeks and $359 million transforming it to accommodate for the F-150's all-aluminum body. During that time 1,100 tractor trailers' worth of robots, conveyor systems and other equipment poured into the facility[1]. More than just the traditional

Audi's A3 body shop. Courtesy of Audi.

resistance spot-welding equipment, Ford brought in an array of other joining technologies, including 500 machines for clinching and adhesive applications, flow-drill screwing, self-piercing riveting and laser welding[2].

While the shift to aluminum "does not necessarily mean more opportunities for lasers," Hurley said the dynamic that over the past few years saw lasers make steady headway in body-in-white (BIW) manufacturing "is at play with aluminum." And just as automakers can choose from a variety of joining technologies for their more aluminum-heavy vehicles, their options are just as numerous when it comes to lasers for welding.

Automotive laser welds are primarily performed by robots, which require fiber optic beam delivery. That eliminates CO_2 lasers as an option for aluminum welding and makes near-infrared, fiber-delivered lasers, such as fiber, disk and diode, ideal choices for these applications, said Eric Stiles, the applications manager for IPG Photonics Midwest. These solid-state lasers are also attractive because highly reflective metals such as aluminum demonstrate greater energy absorption as the wavelength decreases, according to the Aluminum Association's The Aluminum Joining Manual. Solid-state wavelengths fall around 1 µm, compared to 10.6 µm for CO_2 lasers. Fiber and disk lasers are preferable when high beam quality is required, and direct-diode technology is the best solution for brazing and conductive mode welding, said Hurley. Disk lasers have the advantage of being impervious to aluminum's back reflection, which limits the capabilities of fiber lasers during welding, said David Havrilla, manager of products and applications for the Trumpf Laser Technology Center in Plymouth, Mich.

"The design of a fiber laser prevents it from managing the back reflection safely, and although certain measures have been developed to address the fiber laser limitations, this laser will shut down to protect whenever a back reflection occurs," he said.

Additionally, with the introduction of fiber-delivered solid-state lasers with high beam quality, automakers are increasingly using this technology for remote welding. Remote laser welding is a robotic process that enables high-speed and flexible production throughput by using swiveling optics for precise beam positioning[3]. For example, the Ingolstadt, Germany-based Audi AG's A8 features aluminum door panels that are joined by a remote laser welding system. The system, which recently placed second in Germany's Innovation Award Laser Technology, represented the first time an automaker used laser remote welding to join conventional aluminum alloys. It was used for the Audi A8's aluminum doors and resulted in a 53 percent time savings, compared to tactile laser welding[4].

References

1. A. Priddle (Aug. 25, 2014). Ford races to rebuild truck plant for aluminum F-150. *USA Today*.
2. J. Perry (Oct. 1, 2015). Bodyshop of the future. *Automotive Manufacturing Solutions*.
3. The Aluminum Association (2015). *The Aluminum Joining Manual*.
4. Innovation Award Laser Technology 2014. Laser beam remote welding of aluminium for automotive lightweight design. www.innovation-award-laser.org/finalists2016_eberpals. html.

replacement of steel with aluminum, a vehicle's fuel savings can increase by up to 7 percent.

Despite the current strong emphasis on lightweighting, "aluminum in the car business is nothing new," said Jay Baron, president and CEO of the Center for Automotive Research in Ann Arbor, Mich. He pointed out that Ford in 1923 rolled out an aluminum sedan, and the automaker experimented with mixed-material constructions in the 1960s. By the mid-1970s, automobiles had less than 100 pounds of aluminum parts and by 2013 that figure had grown to 350 pounds. Automobiles are expected to have 500 pounds of aluminum by 2025, according to a 2014 study by the Troy, Mich.-based Drucker Worldwide. And in many regards, the U.S. auto industry is just starting to catch up with its European counterpart.

"While Ford was the first to convert a high-volume vehicle to aluminum in North America [with the F-150], Audi and Jaguar Land Rover expanded modern use of aluminum to decrease vehicle weight, improve fuel economy and improve performance more than 20 years ago," said Doug Richman, technical committee chairman of the ATG and vice president of engineering at Kaiser Aluminum in Foothill Ranch, Calif.

What is new, said Baron, is the accelerated replacement of steel with aluminum. And herein lies the opportunity for lasers. Over the past year, for example, Valentin Gapontsev, chairman and CEO of IPG Photonics in Oxford, Mass., has said during quarterly

Figure 3. Ford Motor's F-150 truck in 2015 became the first high-volume vehicle in North America to have an all-aluminum body (inset). Ford used laser welding for the truck's aluminum roof. Courtesy of Ford Motor Co.

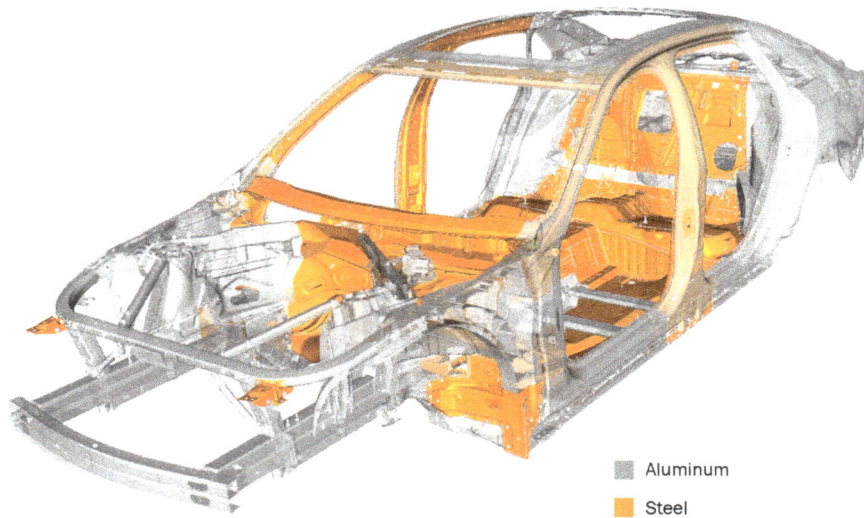

Figure 4. For the Cadillac CT6, GM adopted a mixed-material approach, in which the sedan's body consists of a variety of steel, aluminum and other materials, as compared to Ford Motor Co.'s all-aluminum approach with the F-150. Courtesy of General Motors.

Aluminum
Steel

conference calls that "trends in the auto industry to use high-strength steel and aluminum alloys toward lighter weight automobiles drives increased adoption of fiber lasers."

GM, which primarily uses fiber lasers on the CT6's aluminum, only began using lasers for the welding of aluminum in 2013 with its Chevrolet Corvette Stingray. This vehicle's all-aluminum frame was 57 percent stiffer and 99 pounds lighter than its preceding model[2]. GM's Cadillac Escalade, Chevrolet's Suburban and Tahoe, and GMC's Yukon now also feature an aluminum laser-welded joint, according to Garcia.

Traditionally, automakers' deployment of laser welding had been largely confined to "hang on" parts, such as doors and trunk lids, which attach to the body-in-white, another name for the vehicle shell that gets welded together. However, automakers recently have started to use lasers to weld internal components, such as aluminum fuel filters and steering rods (Figure 5), said David Havrilla, manager of products and applications for the Trumpf Laser Technology Center in Plymouth, Mich. And he expects automakers will soon use this joining technology on aluminum bumpers. Havrilla added that, for electric cars, laser welding of the battery's aluminum casing and the aluminum or copper contacts inside the battery are still in the early stages of development.

Photonics vs. nonphotonics

A significant portion of IPG's fiber lasers are used in car body seam welding/brazing applications. In fact, the company believes the opportunities in the fiber laser welding market are "potentially much more significant" than the cutting market, according to a presentation it gave to investors last May. With its laser seam stepper (LSS), a robotic welding tool that integrates laser welding and clamping to tightly hold work pieces together, IPG is keen on replacing resistance

Figure 5. While traditionally used for the body-in-white, laser welding is increasingly being used for internal aluminum components, such as steering columns (left) and fuel filters (right). Courtesy of Trumpf Inc.

spot welding in several automotive applications. In a recent quarterly conference call, Gapontsev said the LSS "can be used to process several different metals including steel, aluminum, stainless steel — and the strength of the weld is about twice the strength of a traditional resistance spot weld."

While lasers are primarily replacing resistance spot welders, "there hasn't been a wholesale shift in laser usage because of the shift to aluminum," said Tim Hurley, the global automotive segment manager for The Lincoln Electric Co. in Euclid, Ohio. Instead, "laser welding is being employed where resistance spot welding would have been performed on a steel part." He added that aluminum's high conductivity and the marks that resistance welding can leave behind give laser welding important advantages. However, laser welding is sometimes dependent on alloys, some of which are prone to cracking and require filler materials.

"With its high processing speeds, low heat input and resultant low distortion, and overall flexibility of application, this process has become a key assembly technology in the automotive industry during the last two decades," said Richman. "Because laser welding creates excellent quality joints, it is typically used in applications where weld appearance is critical."

While rivets have emerged as a popular joining technology for aluminum parts — with the F-150 featuring over 4,000 of them rather than 7,000 spot welds[3] — Baron said he expects to see automakers rely less heavily on these mechanical fasteners. GM, for example, managed to use about 1,800 fewer rivets and achieve a four-pound weight reduction[4] through its stronger reliance on welding in the CT6.

"Rivets add weight. Laser welding adds no weight," said Baron.

Baron added that lasers' greatest nonphotonics-based aluminum joining technology competition will likely be adhesives, which have become more sophisticated and are better than spot welds. However, Garcia said GM increased its reliance on laser welding, in part, to reduce or replace adhesives and sealers.

Eric Stiles, the applications manager for IPG Photonics Midwest, said laser welding could also replace gas metal arc welding, which is suitable for joining of heavier gauge materials. However, this joining technology's heat input is too high for welding thinner sheet metal. Other opportunities may arise for lasers as automakers adopt part designs that lack flanges or have smaller flanges, according to Stiles, Richman and Havrilla.

Doors opening

"A few automakers are exploring the potential of laser welding to reduce, or eliminate, flanges in sheet metal joints for door, hood and deck designs. In these applications laser welding is used to replace traditional formed seams joining inner and outer panels in sheet metal assemblies. If successful, this may be an application with wider potential for laser welding deployment in the future," said Richman.

Parts such as doors typically have weld flanges, upon which pressure is applied during the spot-welding process as pieces of metal are joined by an electric current. With laser welding, for example, the door window flange of the CT6 is 6 mm, compared to 10 mm to 15 mm with a typical spot-welded door window flange. And such small reductions, when repeated throughout the body, lead to significant weight reductions. Garcia said the linear stitch shape of laser welding allows for smaller flanges, whereas spot welds have a circular 5- to 7-mm shape.

Illustrative of the weight savings afforded by the combination of aluminum parts and laser welding, the Valley City, Ohio-based Shiloh Industries in 2014 unveiled the industry's lightest, mass-produced aluminum door blanks, which consist of sheets of varying thickness and grade. Shiloh's aluminum laser-welded door-inner blanks weighed 26.8 pounds, compared to 35.5 pounds for four aluminum, monolithic door-inner blanks of similar geometry. And while the aluminum laser welded door-inner blanks yielded a nearly nine-pound weight reduction per vehicle, the savings were even greater — 58 pounds — when compared to steel monolithic door blanks.

This is the time for industrial lasers to cut through the competition or get left behind by alternatives such as adhesives and spot welding.

Overcoming costs

The costs of aluminum and lasers have long hampered the adoption of aluminum laser welding in the automotive industry. While Havrilla points out that over the past 10 years the cost of a 4-kW laser has fallen 66.7 percent to $200,000, the Steel Market Development Institute, citing a Massachusetts Institute of Technology study, notes that steel is two to three times cheaper than aluminum. And aluminum's cost is not dropping as robustly as that of lasers. Between 2006 and 2015, the annual average price per metric ton of aluminum has only dropped 12 percent to $1,665, according to the statistics firm Statista.

"If aluminum truly takes off in [the] automotive sector, it will drive the cost of aluminum down while also increasing process knowledge. With greater understanding of the process and increased affordability, all fabricators are more likely to consider aluminum as an easier and more desirable material to work with," said Havrilla.

Baron said that, with automotive business models trending toward the addition of more technology, "lasers are positioned for more growth." Garcia added that GM's usage of lasers "has also grown as we become more confident with the technology." And given how the automotive and aerospace industries customarily serve as trendsetters, laser makers anticipate some general manufacturers to jump on the lightweighting train. While few other industries are as weight-sensitive as these two, many can benefit from the shipping relief aluminum can provide.

Hurley at Lincoln Electric said the agriculture industry is looking into aluminum to get "more bushels to the gallon of diesel." Havrilla said extruded aluminum might also be attractive to window frame manufacturers because it could help reduce shipping costs, when compared to steel frames.

"Aluminum has many benefits, from lightweighting to corrosion resistance, so as usage of aluminum increases in the automotive industry, there will certainly be some cross-pollination with other industries," said IPG's Stiles.

References

1. D. Sherman (May 20, 2016). How GM is saving serious weight in its vehicles now — and in the future. *Car and Driver.*
2. A. Luft (Jan. 26, 2013). Deep dive: The light, yet stiff frame of the 2014 Chevy Corvette Stingray. *GM Authority.*
3. L. Brooke (Jan. 10, 2014). The F-150's aluminum diet. *The New York Times.*
4. R. Truett (May 24, 2016). GM lightweighting strategy starts to bear fruit. *Automotive News.*

New Milestone in Laser Bonding

An innovative process involving a high-power, short-pulse laser is the key to creating strong metal and plastic joints for the automotive and aerospace industries.

BY FLORIAN KIEFER, TRUMPF INC.

The demand for lightweight construction and economic efficiency has led to R&D efforts in metal-plastic combinations. Although primarily driven by the automotive industry, this has become an increasingly important topic for other industries as well. By using a high-power short-pulse laser it is possible to establish a simple joint with a very high loadability. Design engineers are therefore gaining access to the completely new and innovative application possibilities of hybrid assemblies.

Prospects of hybrid assembly

Metal is the most common material chosen for the production of load-bearing parts in the automotive and other industries, yet the continuous improvement of plastics in the last decades offers new possibilities for its use. By replacing metal parts with plastics, manufacturers can reduce the weight and costs of manufacturing. Using the automotive sector as an example, this development not only eliminates vehicle weight while reducing gas consumption, it also reduces the material and production costs overall. The crucial issue has been creating an acceptable joint between the two materials, metal and plastic.

Image taken during the structuring process. The structured material appears at the left of the laser beam. All images courtesy of Trumpf Inc.

There are primarily two different types of joining technologies for combining metal and plastic. The plastic can either be molded directly onto the metal part in a process called in-mold assembly (IMA), or the molded plastic part and the metal part can be created separately and then bonded together during an additional processing step referred to as post-mold assembly (PMA). During the IMA process, the metal part is placed in the molding tool and then the form-fit joint is created within the molding process.

Drawbacks to traditional approaches

The IMA process leads to broad constraints in design flexibility since the metal part has to be adapted to the molding process. The design of the molding tool is also significantly more sophisticated because IMA requires a sealed joining process. These reasons are likely why PMA processes are more commonly used. However with PMA, manufacturers still struggle to meet the mechanical demands and often need additional connection elements to join the materials sufficiently. Blind riveting, punch riveting and screwing are the most common processes that require

The high-power short-pulse laser is very stable and scalable due to its architecture, which is based on a disk laser, combined with cavity dumping technology. With this technology, the pulse duration remains at 30 ns throughout the complete range of the frequency. It also allows for the average power to remain stable at 850 W with a maximum of 80 mJ.

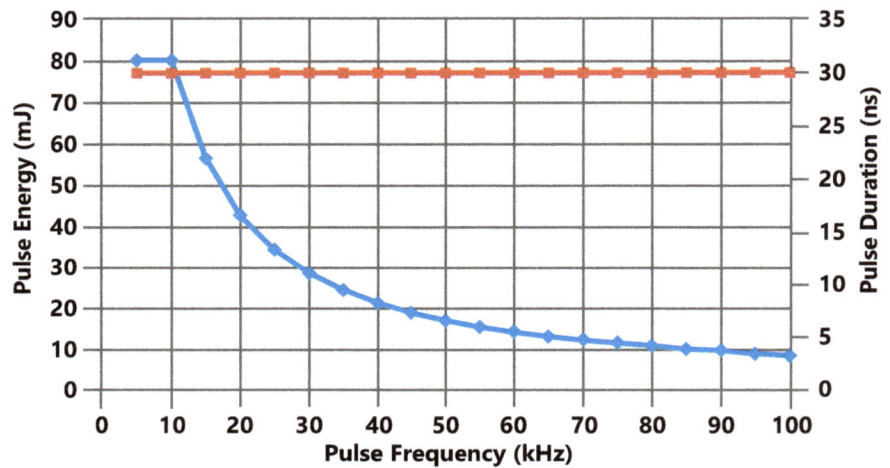

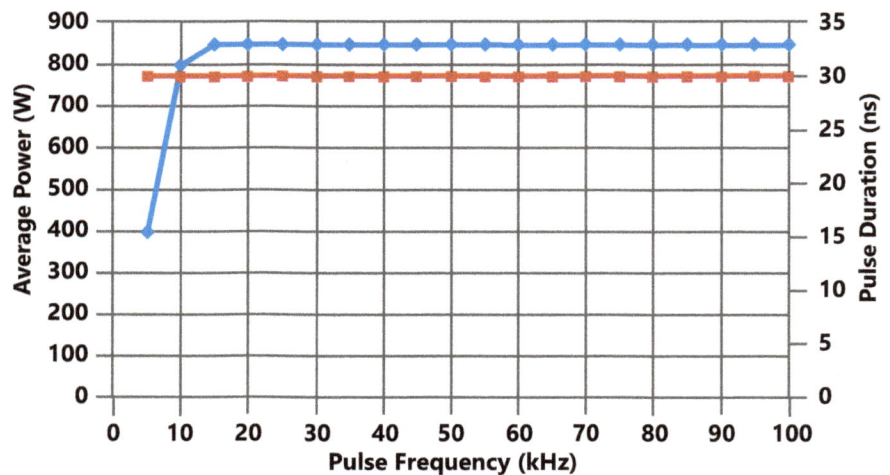

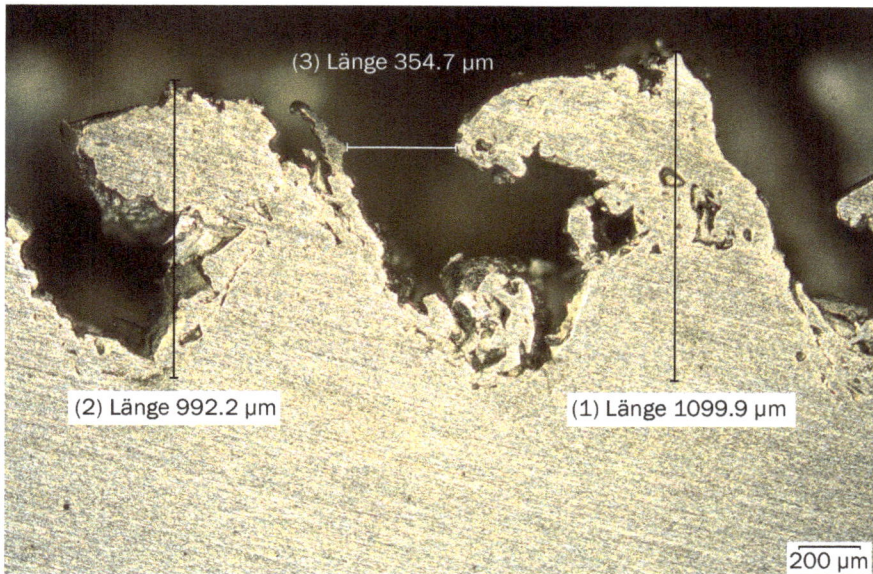

A cross section of an aluminum part, with dimensions indicated, after it is structured with the laser.

this additional connection. The concern here is that this added connection is the weak point between the elements and, therefore, the point of failure. Furthermore, when a gas or watertight combination is required, sealing and additional efforts become necessary, which only contributes to the time and resources required for joining the materials. Another method of combination is adherence. In addition to the adhesive, usually a preparation of the adherend is essential to reach the mechanical demands. This solution also adds an additional processing step. A PMA method that does not require additional connection elements is, for example, riveting technology. For this, a plastic molded pin is deformed to connect metal and plastic parts. To strengthen this connection method, extremely precise positioning of the holes in the metal to the pins in the plastic is needed. Additionally, these connections tend to become loose at high temperature changes.

Innovation in short-pulse lasers

The latest advancements in laser technology have enabled a completely new way of joining metal and plastic. Using a high-power nanosecond laser, it is possible to produce a metal-plastic bond that is not only highly loadable but one that also offers a wide range of mechanical advantages such as a fit that is watertight, gastight and resistant to temperature changes. Neither additional connecting elements nor complex component handling is needed. The process consists of two steps: laser structuring of the metal and joining of the metal and the plastic part.

A high-power short-pulse laser is a beam source that emits short light pulses with high energy. For laser structuring of the metal-plastic joints, a nanosecond laser is used. Light travels the distance from the Earth to the moon within one second. For comparison, the length of a nanosecond pulsed laser beam is about one foot. This short-pulse length in combination with the high-pulse energy leads to pulse peak powers of more than 2.6 megawatts, which is similar to the power of an average wind turbine. At the same time, the average output power of the laser ranges from a few watts up to hundreds of watts. The extreme difference between high peak power and average output power is possible due to a repetition rate of more than 100 kHz.

The design of the high-power short-pulse laser is based on a disk laser architecture. This grants high efficiency paired with low maintenance costs and high reliability. These solid-state disk lasers are common for industrial applications, and are most often used for laser cutting, welding and drilling. The high-power short-pulse laser is widely used in the solar industry or for laser cleaning of various parts. For these types of processes, high-pulse peak powers are particularly advantageous in removing coatings or dirt from the base material without damaging it.

Fast-moving beam

When using a high-power short-pulse laser for joining metal and plastics, the first processing step generates a microstructure on the surface of the metal. Short laser pulses with high-pulse peak powers create high energy densities that lead to both a partial evaporation and a partial melting of the material. The short pulse combined the short energy transfer leads to an immediate freezing of the molten mass. The metal freezes in barb-shaped edge beads generated by high gas pressure at the position of the laser spot.

Such a process is only possible if the laser beam is moved over the metal part at a fast enough speed, which is realized using scanner optics. Within the scanner op-

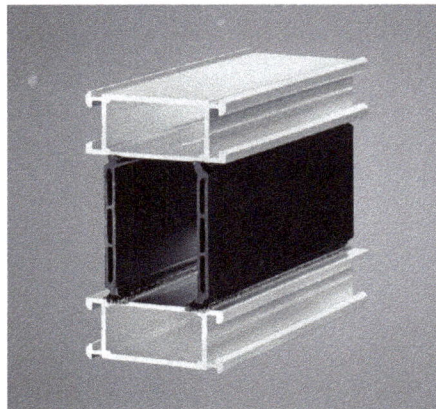

A metal-plastic bond is generated using a high-power nanosecond laser designed by Trumpf.

tics, the laser beam hits two actuator-controlled mirrors. The slightest tilting of the mirrors moves the laser beam at a very fast rate along a distance of some millimeters. Using a special scanner movement and the right set of laser parameters, it is possible to create structure depths ranging from micrometers to 10ths of a millimeter. The resulting structure consists of many small undercuts and, depending on the demands, the generated structure can be either direction-dependent or direction-independent. By using laser pulses in the nanosecond range, the material temperature always remains low enough to avoid structural changes to the material itself. Thus, the strength of the base materials remains unchanged. Altering the energy of a pulse within the same setup, the treatment of steel, aluminum, copper and titanium is possible. In principle, every metal can be treated as long as the energy of the laser is sufficient to generate the melting phase of the metal. Furthermore, controlling the structure depth allows a fabricator to fine-tune the process velocity or the strength of the connection, which opens up a range of possibilities for applications.

A microstructure is created on the surface of the metal through partial evaporation and partial melting of the material by the laser.

Joining process

As a second step, the structured metal is joined with the plastic. Therefore, the plastic has to melt into the prepared metal structure. The joint can either be done with the IMA-process or with the separate PMA-process. Using the IMA, the metal part is placed inside the molding tool. The plastic is injected directly onto the undercut profiled metal and the bond is created. For the PMA process, the metal part is heated locally, for example through induction or by using a laser. Considering the properties of the plastic, the correct temperature for the metal is chosen. The temperature must be high enough to melt the plastic but low enough to not cause chemical decomposition. The prepared plastic part is then pressed against the hot metal. The plastic melts and flows into the undercuts. A strong connection between the metal and plastic is formed after the plastic is refrozen. With the ability to regulate the structure depth, this process can be successfully employed for joining nearly every type of plastic. In addition, a structure with coarser roughness also enables fiber-reinforced plastic to be used. The IMA process additionally allows for use of a duroplast.

Contact-free laser treatment

With the laser structuring process, the resulting joint resembles a zipper in that the metal and plastic parts are interlocked through small barb-shaped undercuts that are in the micrometer range. This joining method grants high cohesion with high flexibility. In addition, impressively large forces can be exerted in any direction on the complete metal-plastic part due to the structuring of anisotropic undercuts. The result is a joint that can exceed the strength of the untreated plastic. With minimal distance between connective points, the difference in thermal expansion of the metal and plastic is nearly negligible. As a result, the joint is extremely resistant against varying temperatures as well as against dynamic stress. It is also possible to design a water- and gastight metal-plastic joint.

The contact-free laser treatment of the metal enables manufacturers to structure flat, round, or even more complex parts of any form. The continuous bond created between the two materials and the resulting high loadability of the joint enables fabricators to join smaller areas and thereby achieve high joining velocities. The process velocity of a 5-mm-wide area is greater than 3 m/min, which is comparable to the velocities of a laser-welding process. After setting the process-specific parameters to perform at its optimum level, the joining process is both fast and easy to reproduce. The high degree of flexibility when it comes to choosing a suitable material, combined with the steadily increasing areas of usage for plastics, make this laser joining process applicable for any industry.

Laser structuring applications can easily be found from the aerospace and automotive sectors to energy engineering — virtually any place where metal meets plastic. Furthermore, laser structuring is not only a convenient and innovative solution for joining metal and plastic; it can also be used to increase friction between two components. Laser structuring is especially well-suited for this application since the various materials can all be structured using a single laser. By adjusting the structure depth, the friction can also be fine-tuned to a specific value, depending on the desired result. Laser structuring can also be used as a preparation step for coatings. With this technique to increase the contact surface between the materials, the bond between a metal part and a ceramics coating can be made significantly stronger. As an added benefit, the heat transfer between base material and coating also improved. The possibilities of laser structuring are endless. For many parts, completely new perspectives will arise as the mechanical properties of metal-plastic joints and additional applications for laser structuring are further understood and applied in manufacturing.

Meet the author

Florian Kiefer is a senior application engineer at Trumpf Inc. in Farmington, Conn. In this role, he supports Trumpf's short- and ultrashort-pulse solid-state laser product portfolio.

Bringing a Third Dimension to Laser Materials Processing

Using a tunable lens enables 3D laser processing without any mechanically moving parts.

BY DAVID STADLER AND JÖRG WERTLI, OPTOTUNE SWITZERLAND, AG

The use of laser light for materials processing has become a standard practice with widespread use in industry. The basic idea is to focus a laser beam onto a fixed working plane and create a spot of maximal intensity. At this position the laser processes the surface to drill, engrave or mark the work piece. Fixed glass optics, in combination with movable galvo mirrors, provide for the scanning of the whole working plane in horizontal X and Y directions. A typical scan field measures 100×100 mm. However this is restricted to one specific working distance (2D), whereas objects generally are three-dimensional (Figure 1).

In recent years, laser processing applications have sought to overcome this limitation and access the third dimension (Z-axis), typically by using mechanically moving Z-stages. In a complementary approach, Optotune has recently developed a tunable lens dedicated to laser processing, shown on the left of Figure 2. The lens has a 10 mm clear aperture and the optical coatings are optimized for 1064 nm laser wavelength. This enables 3D laser processing without any mechanically moving parts. Consequently, the time scale of focus change is very fast, in the range of several milliseconds.

Figure 1. Difference between 2D and 3D laser processing: The laser spot has to be adjusted in Z-direction to process the 3D work piece, as shown on the right.

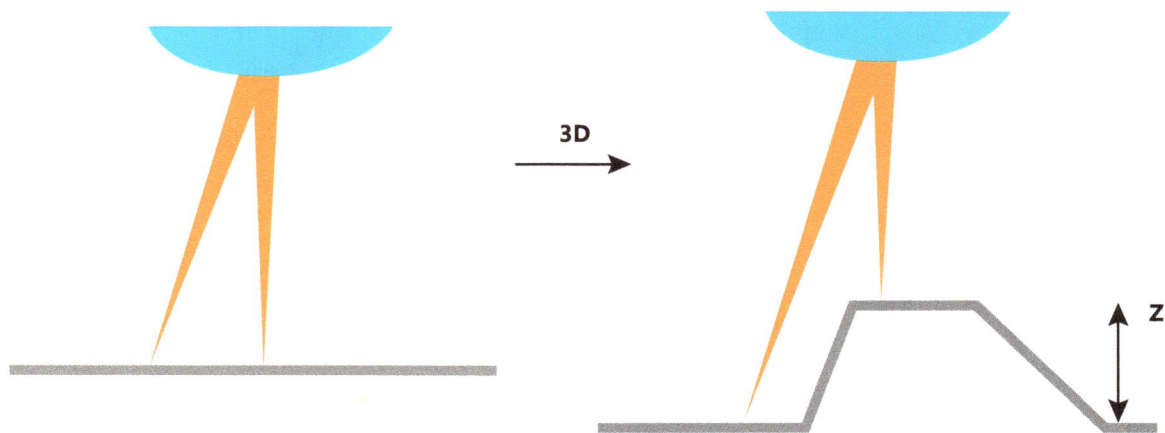

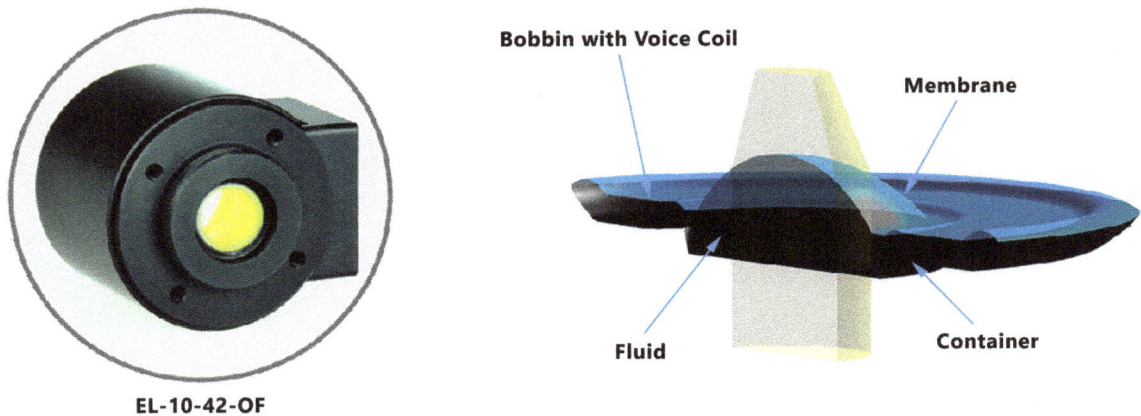

Bobbin with Voice Coil

Membrane

Fluid

Container

EL-10-42-OF

Figure 2. Tunable lens technology: The left side shows an image of the lens module and the right schematic illustrates the working principle of the tunable lens.

Based on tunable polymer lens technology, the Z-position of the focus spot is changed by directly tuning the focal length of the tunable lens from −500 mm to +500 mm. The basic working principle of the tunable lens is illustrated on the left side of Figure 2. It consists of a container that is filled with an optical fluid and sealed off with an elastic polymer membrane. The lens core has an electromagnetic actuator that is used to exert pressure on the container, changing the deflection of the lens. This allows for controlling the focal length by changing the electrical current in the surrounding coils.

High focus stability and repeatability are essential parameters to guarantee consistent processing quality. If the laser is not focused on the working piece, the peak intensity quickly drops due to the increased spot size. Focus drift mainly originates from temperature effects. The volume of the fluid in the lens core slightly changes with temperature, leading to a change of deflection and focal length. That is why the technology presented has an integrated optical feedback that measures the deflection of the lens in-situ by detecting the probe light of an LED on highly sensitive photo diodes. Consequently, a position stability well below the Rayleigh length of a typical setup, the relevant figure of merit with which to compare, is possible. To further improve the long-term focus stability, the tunable lens is operated at a fixed working temperature of 47 °C. The internal heating element requires approximately 5 minutes to heat the lens after startup. Thermal isolation between external parts and the lens is ensured by a Teflon ring placed in between.

A controller card provides all the necessary analog and digital electronics to control the tunable lens. The hardware is optimized for the specific needs in laser processing applications, such as a minimal sized board and convenient interfacing to external electronics. To control the focal length of the tunable lens through the controller card, an analog voltage signal between 0 and 5 V is used. Through calibration, this voltage range has to be mapped to the tuning range of the laser spot in the Z-direction. This typically is done through a look-up table that maps a certain voltage to the corresponding Z-position (mm). This is, to a very good approximation, a linear relation. Furthermore, several digital signals for error handling are available, such as a "lens ready" signal. This can be used to trigger the laser at the right moment when the lens has reached its set point.

The small dimension of the technology presented allows for a very compact integration. A typical laser processing system provides enough space between the laser head and the galvo scanner for the tunable lens to be installed. As an application example, a table-top laser marking system has been set up. As a laser source, a 20-W fiber laser with a wavelength of 1064 nm is used. The beam has a diameter of 6 mm when entering the tunable lens. The galvo unit is a scan head with 10-mm sized mirrors. It is then guaranteed that the beam will not be clipped by the mirrors.

Two principle schemes of integration have been realized with the demonstration setup (Figure 3). On the left, the 2.5D approach with an f-Theta lens of 160-mm focal length is shown. In this configuration the f-Theta lens provides the field flattening at each Z-position while the Z-position of the laser spot is shifted by the tunable lens. The central working distance is at 191 mm and the resulting Z-tuning range of the system is about 100 mm.

It is important to note that the field size increases for longer working distance simply due to geometry. The tunable lens is placed in the optical path before the scan head. The exact position is not very critical because the distance between the f-Theta lens and the tunable lens has only a very small influence on the available Z-tuning range. However, it is recommended to place the tunable lens as close as possible at the entrance aperture of the scan head.

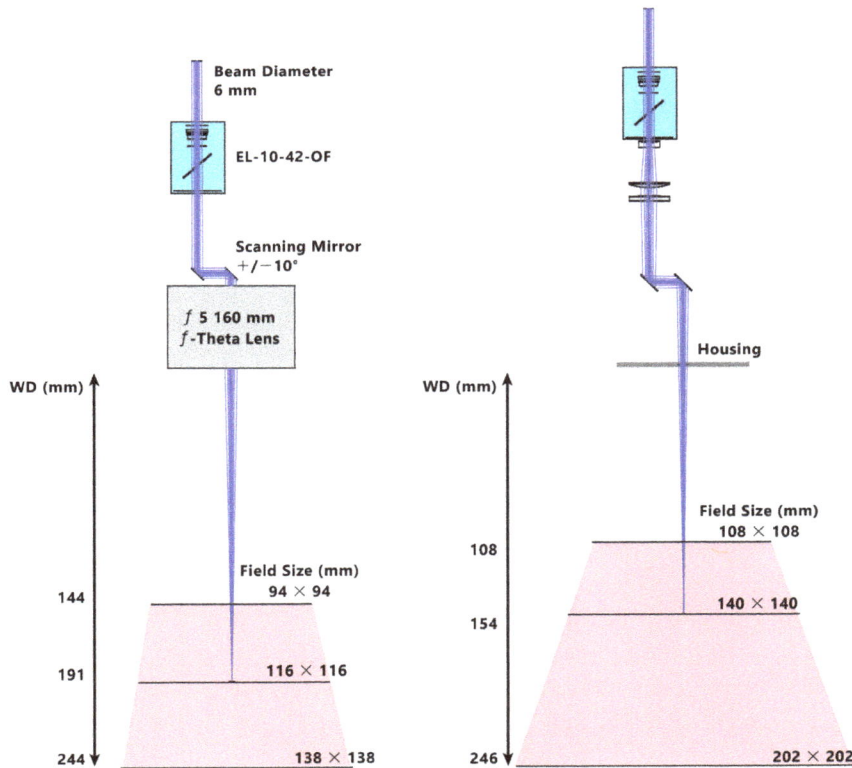

Figure 3. Two principles of integration: The left schematic shows the 2.5D setup with an f-Theta lens of f=160 mm focal length. The working distance ranges from 144 to 244 mm and the field size changes from 94 × 94 mm to 138 to 138 mm, set by the mirror angle of 20°. The right schematic illustrates a possible setup for f-Theta free 3D marking. The tuning range and field sizes are larger compared to the system with f-Theta lens.

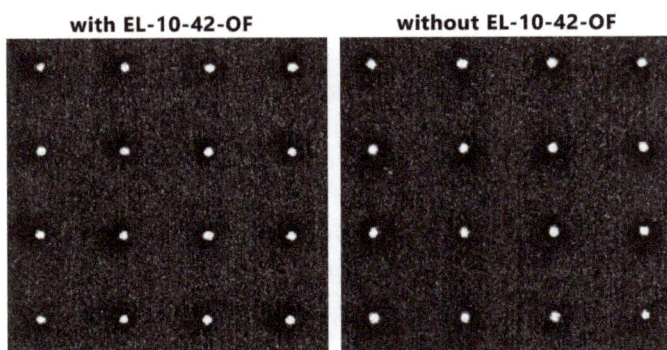

Figure 4. Comparison of marking quality with and without the tunable lens on anodized aluminum. When analyzing the 4 × 4 dot matrix under an 8× microscope, no significant difference is visible.

The 2.5D system is a practical and simple solution when marking on horizontal layers at different heights. The time needed to jump from one layer to the next is a relevant time scale to quantify the marking speed. For a 10 to 90 percent step at the input corresponding to 0.5 and 4.5 V respectively, it requires approximately 12 ms to reach the new set point. The response time can be divided into two parts. First, there is a constant delay of about 3.5 ms due to the finite update rate of the analog-to-digital converters of the controller card. Then there is the rise time of about 8.5 ms to reach the new set point. For small steps the response time decreases to about 6 ms.

The integration for 3D marking is shown on the right of Figure 3. This approach is technically more appealing, but it is also more challenging. Field-flattening and Z-shift is accomplished by the tunable lens, making the f-Theta lens unnecessary. The optical layout presented here is based on off-the-shelf lenses. Essentially, the tunable lens is followed by a beam expander and a focusing lens that creates the laser spot at a distance of 154 mm measured from the housing aperture. Due to beam expansion, the spot size remains roughly the same as for the 2.5D system with f-Theta lens. However, the Z-tuning range and field size is larger. The implementation of 3D marking with an f-Theta lens is also possible, but this approach lacks the advantage of reducing costs.

For highest 3D marking speed the different dynamical behavior of the tunable lens and the galvo scanner must be considered to guarantee synchronized operation. The tunable lens has a resonance frequency of about 200 Hz, resulting in a lower bandwidth than typical galvo scanners that can have a bandwidth of several kHz or more. This poses a challenge on both software and hardware integration to drive the combined 3D system at its physical limit. At present, on a surface that is tilted by 45°, a 3D marking speed of 700 mm/s can be achieved with the demonstration setup, limited both by software and the controller card. A desirable solution would be to have a fully digital controller board that controls all three axes at the same time or a separate controller with higher update rate and a digital interface. Such a solution is expected to increase the processing speed by factors, achieving typical speeds of standard 2D laser marking systems.

While speed is one aspect regarding the applicability of the tunable lens, the achievable processing quality is also important. This requires good optical quality of the utilized lenses, which can be quantified by the root mean squared (rms) wavefront error. The technology presented has a value of typically 0.15 of the laser wavelength, which is sufficient for many processing applications. With the demonstration setup Optotune marked a matrix of four-by-four dots on a plane. Each dot has a diameter of about 80 µm. The material used was anodized aluminum. Figure 4 compares the two situations with and without the tunable lens in the beam path. When analyzing the result under a microscope with 8× magnification, no significant difference is visible.

In a next step the optical design, software and control electronics have to be considered to achieve optimal performance in full 3D laser processing applications. First attempts in that direction are already initiated.

Meet the authors

David Stadler is an application engineer with Optotune Switzerland AG in Dietikon, Switzerland. Jörg Wertli is the head of marketing with Optotune.

Remote Laser Welding in Industrial Applications

Laser manufacturing requires time-tested, reliable solutions.

BY DR. KLAUS KRASTEL, TRUMPF LASER UND SYSTEMTECHNIK GMBH, DAVID HAVRILLA, TRUMPF INC., AND HOLGER SCHLUETER, TRUMPF PHOTONICS

The market for industrial laser systems is very mature and has been growing steadily for almost 30 years. In 1979, the first two-dimensional laser-cutting systems were introduced as tools for the fabrication industry. Today, this market has expanded from cutting to welding and from two- to three-dimensional applications.

Customers have grown to expect industrial solutions and are not typically concerned with the technology behind the solutions. They expect a semi- or fully automated tool with an uptime greater than 95 percent, and proven processes and operating procedures. They are not focused on laser technology but rather on overall system performance.

This need for process and system performance has created a marketplace that is dominated by solution providers. A solution must include the process — usually

Figure 1. This laser machining center has a cartesian five-axis motion system and a 6-kW CO_2 laser. These are labeled X,Y,Z for the linear axes and B,C for the two rotational axes.

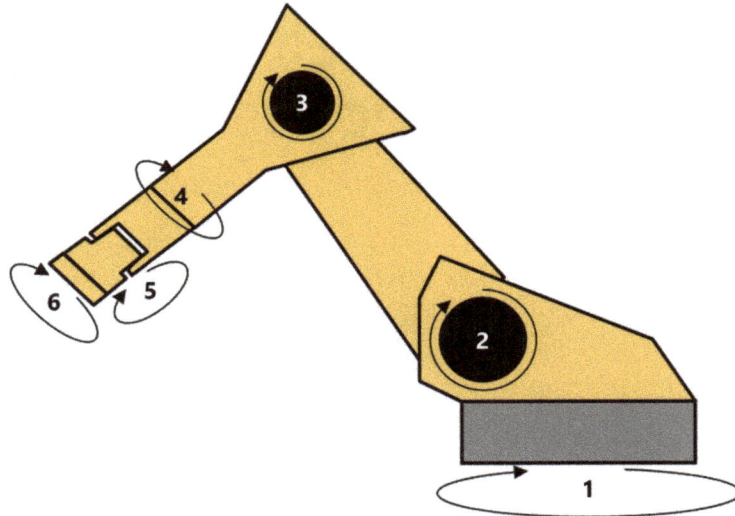

Figure 2. A robot has six axes of rotational motion.

encoded in technology tables within the machine — the motion system, beam delivery, programming software and the laser source. Typically, the laser source accounts for 25 to 30 percent of the value of such a laser-processing solution.

New beam delivery

Two-dimensional laser-cutting systems generated more then $4 billion in revenues in 2005. Interestingly, the machines are typically sold to fabricators with 10 to 20 employees. This means that the 2D laser-cutting process has become so standardized that a small job shop can purchase the machines without undertaking extensive operator training or possessing complicated programming experience.

Three-dimensional laser cutting and welding systems require more know-how in the programming of the tool path, as do five-axis milling machines and spot-welding robots. Similar to a five-axis milling machine, modern 3D laser cutting and welding systems use cartesian five-axis motion systems and achieve tool path accuracies of under 1/1000 of an inch (Figure 1).

A cartesian five-axis motion system is characterized by three linear axes of motion that are perpendicular to each other and two rotary axes that enable positioning of the welding or cutting head normal to any desired surface on the workpiece.

In contrast, a robot typically has six axes of motion, all rotational (Figure 2). To achieve a movement and orientation in cartesian space, the robot's controller must solve a complex equation in real time, transforming the coordinates XYZ and the orientation of the tool tip into the angles of the six rotational axes. This transformation can sometimes have more than one solution, and the robot's controller must decide which one is better suited to a given situation.

By solving this equation, a robot can situate a beam in the same position and with the same orientation as a cartesian five-axis motion system. However, as soon as it needs to follow a path (during welding, for instance), the inertia of some and the motion of other individual axes makes them strongly interfere with each other. It then becomes impossible to solve the extended equation — which takes into account angular overshoot of the rotational axes because of varying inertia — with high accuracy. The result is that the path accuracy of a robot is fundamentally less than that of a cartesian five-axis motion system.

In general, robot-based laser processing solutions are limited by three inadequacies:

1. The path accuracy of a robot is not adequate for modern manufacturing requirements. For instance, circles and similar geometric shapes typically do not meet customer demands.

2. A robot heats up during operation and cannot achieve the same positioning accuracy over long manufacturing periods as can a cartesian five-axis motion system.

3. Positioning and processing speeds of robots are well below those of modern cartesian five-axis systems.

Cartesian five-axis motion systems are mainly used in cutting applications, where superior speed, and position and path accuracy are required. However, they tend to be substantially more expensive than robot-based systems, so the total addressable market is limited to high-value applications.

For high-volume, medium-value welding applications — for instance, most automotive welding applications — the first two deficiencies of robot-based systems are less relevant and have allowed for a laser market of a few hundred million dollars for robot-based 3D laser welding systems.

In this market segment, laser powers have been steadily increasing over the past few years (from around 3 kW in 1998 to 8 kW in 2007), and this has pushed the welding process speeds to values where the robot's No. 3 deficiency — its speed (see above) — becomes the most important obstacle to increasing throughput and productivity.

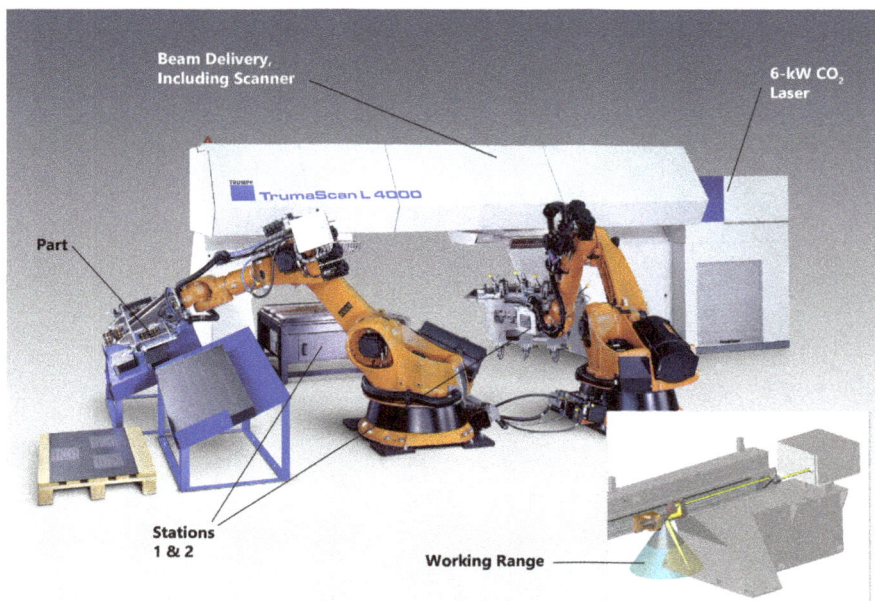

Figure 3. This CO_2 laser remote welding system has a 6-kW laser and two workstations.

These stumbling blocks have led to the development of a new beam-delivery solution for welding applications. Recently, two techniques were realized that eliminate the need for the fast and accurate mechanical axis of a cartesian five-axis machine and that utilize high-power scanning optics to position the beam within the work envelope. The industry has termed these beam scanning systems "remote laser welding" stations.

These systems greatly reduce the time it takes to position the beam between two welds, increasing the beam-on time, and also allow maximum process speeds, even when enabling complex path geometries with small radii.

Remote laser welding

The first new approach is a remote system for a 6-kW CO_2 laser (Figure 3). The system consists of the laser source on the far right and the beam delivery and scanning system extending over the work envelope. The configuration shown has two separate work zones, enabling better productivity by alternating laser welding and part loading and unloading between the two work zones.

Part loading and unloading can be accomplished via different methods — dial table, robot, linear transfer. The system shown in Figure 3 uses robots to pick up the parts and to hold them in place under the scanning system. During processing, the laser beam focus is adjusted up and down by moving the focusing mirror (focal length: 1.5 m), while the XY positioning is achieved by deflecting the now-converging beam with a mirror on a kinematic mount that allows deflection of up to 30° in both directions.

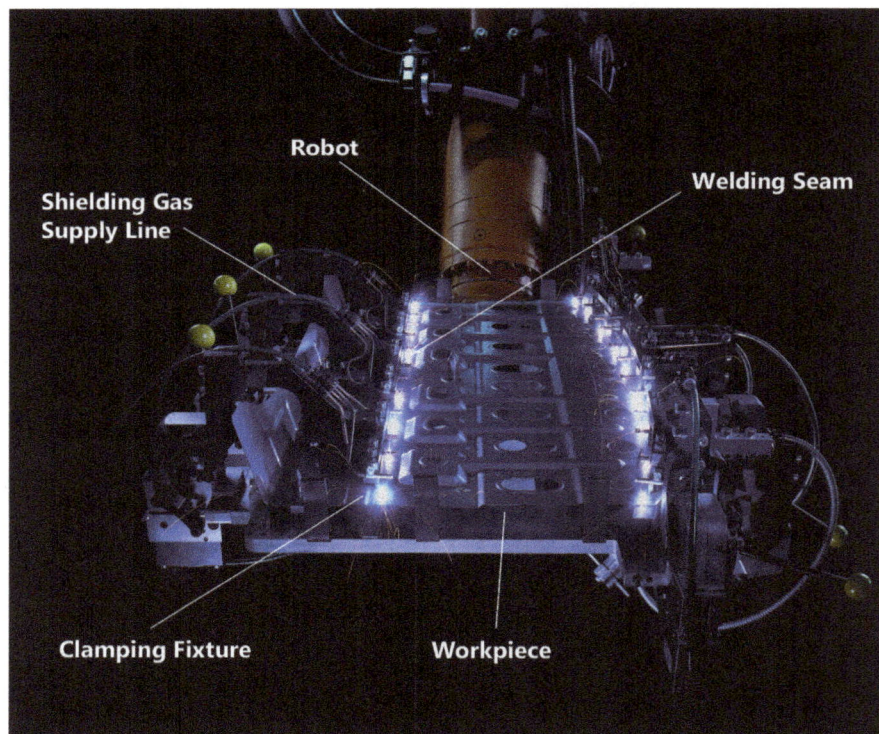

Figure 4. Specialized fixturing holds the workpiece for laser remote welding with a 6-kW CO_2 laser.

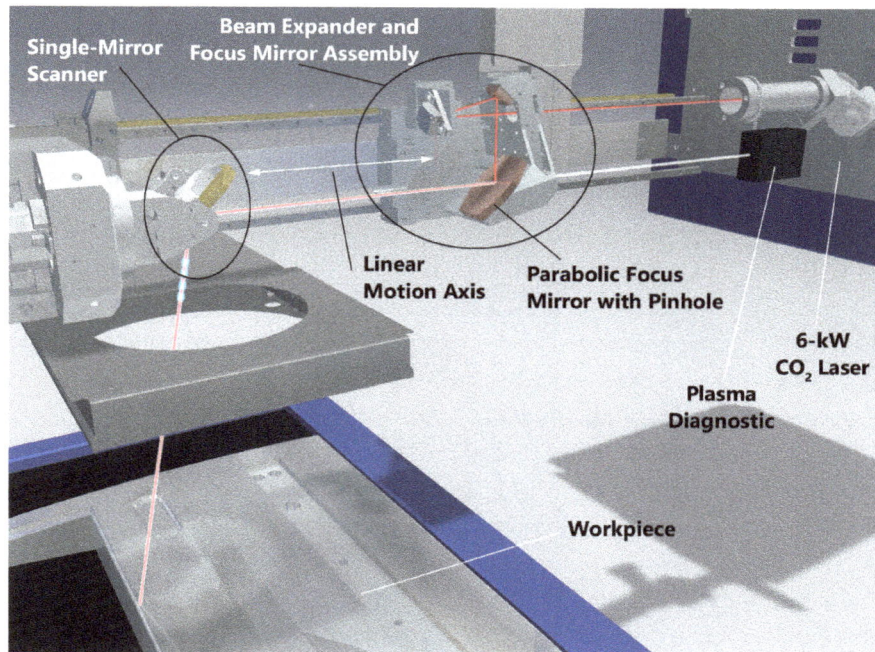

Figure 5. The internal optics of this remote-welding station allow the beam power to be delivered anywhere within the work envelope.

Figure 4 shows the workpiece held in place by the robot, while a long-exposure photograph allows visualization of the laser beam path along the weld seams. An important aspect of remote laser welding is that the welding shielding gas must be supplied through the part fixturing. Clearly visible in this picture are the gas supply hoses running past the clamping mechanism with the yellow knobs. This is a departure from highly flexible no-tooling laser processing solutions, as it necessitates dedicated fixtures for remote laser welding. However, this is a price the industry is willing to pay for the increased productivity provided by a remote laser welding system.

Figure 5 shows a schematic of the normally enclosed beam path. The laser beam exits the 6-kW CO_2 laser parallel to a linear-motion axis. Two separate optical assemblies can move parallel to the linear-motion axis:

1. A beam-expander and focus-mirror assembly.
2. A single-mirror scanner assembly.

The beam-expander and focus-mirror assembly consists of three mirrors: a parabolic focus mirror and two that form a beam-expanding telescope. The parabolic focus mirror is crucial for consistent process results. Although a lens could potentially achieve the same focal length and spot diameter as a parabolic mirror, transmissive optics experience major thermally induced focal shift when the 10.6-μm laser power changes. It is typically not possible to compensate for such focal shift, which also depends greatly on the age of the optics and on the degree of contamination.

The parabolic mirror solution has another advantage: It allows for gathering the process light through a center hole for plasma diagnostics.

After passing through the beam expander and focus mirror assembly, the convergent beam hits a single-mirror scanner that deflects the beam in the X- and Y-directions. At the beginning of an operation, the single-mirror scanner assembly moves to a predetermined position over one of the workstations. (Figure 3 shows a system with two stations, but up to four stations are possible.) It can stay fixed, while the parabolic focus mirror adjusts to provide very fast Z-axis motion of the laser beam focus. When angular motion of the single-mirror scanner is added to the focal-length adjustment, the system can address any position within the work envelope.

The second approach to remote laser welding is a system that — unlike the variable-focus one discussed above — relies on coarse and fine position adjustments. The coarse adjustment is performed by the robot arm, and the fine adjustment is achieved with the programmable focus optics.

The system shown in Figure 6 uses a 4-kW thin-disk laser, but powers up to 8 kW will soon be commercially available. A beam switch inside the laser housing can switch the laser output among several workstations, and the beam is delivered to the stations via an optical fiber.

Figure 6. A laser network enables time-sharing in a remote welding application.

Although reflective optics were necessitated by the 10.6-µm radiation of the CO_2 laser in the previous example, transmissive optics are well suited for the 1.06-µm

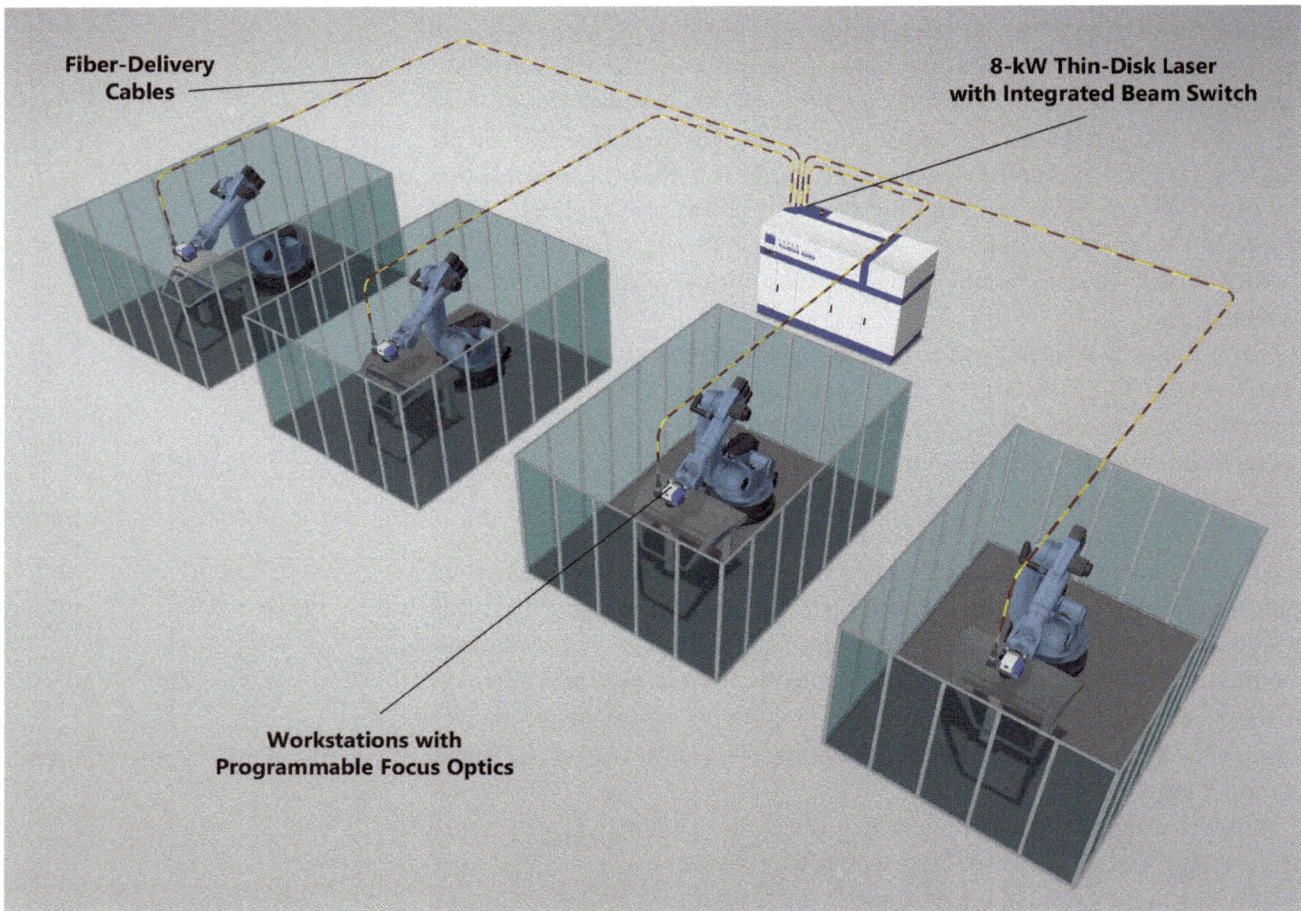

Fiber-Delivery Cables

8-kW Thin-Disk Laser with Integrated Beam Switch

Workstations with Programmable Focus Optics

radiation from the thin-disk laser because the absorption of the radiation in the quartz optics is many orders of magnitude lower. Therefore, there is no thermally induced focal shift, and it is even possible to use beamsplitters to enable plasma diagnostics.

The coarse movement of the robot arm must be synchronized with the fine movement of the programmable focus optics. The position-feedback loop of a robot is usually updated at 500 Hz, which is far too slow for accurate synchronization between the robot motion and the motion of the scanner in the programmable focus optics.

Therefore, a robot-based remote laser welding system typically includes an enhanced robot/scanner control that allows better path accuracies and also enables the concurrent movement of the robot and the scanner, which is crucial to efficient production. The system would be far less efficient if the two motions had to take place sequentially.

The beam switches that are integrated into modern solid-state lasers — such as in the networked stations shown in Figure 6 — permit two major improvements over single-laser workstations:

1. The beam of one laser can be time-shared between different stations.

2. In the case of scheduled maintenance, a second laser can provide redundancy for the laser that is out of commission.

This switching capability has been a huge part of the success of the solid-state lasers in automotive welding.

Looking to the future

Modern remote laser welding systems enable new welding applications because of drastically improved productivity and speed. In recent years, stationary systems based on CO_2 lasers and robot-controlled systems using solid-state lasers have been introduced into the industrial marketplace. They provide excellent performance and flexibility, while providing the possibility of online quality control through optical process monitor capability. Remote laser welding is the latest in a series of successful industrial laser solutions that have revolutionized sheet metal processing over the past 30 years.

In the future, this technology will lay the groundwork for further innovations in cutting and welding.

Meet the authors

Klaus Krastel is an R&D engineer for application development Laser and Systems at Trumpf Laser und Systemtechnik GmbH in Ditzingen, Germany, and is responsible for CO_2-remote welding.

David Havrilla is the YAG product manager at Trumpf Inc. in Plymouth, Mich.

Holger Schlueter is general manager and vice president of Trumpf Photonics in Cranbury, N.J.

Lasers in the Manufacturing of LEDs

Cost-effective industrial tools are helping manufacturers meet increased demands.

BY MARCO MENDES AND JEFFREY P. SERCEL
J.P. SERCEL ASSOCIATES INC.

In today's society, there is a continuous need for devices with lower energy consumption and higher efficiency. Light-emitting diodes (LEDs) are expected to see a 61 percent rise in worldwide demand in 2010, according to Barry Young of IMS Research, due in large part to the mobile handset. The market for large backlit LED TVs is rapidly expanding, and LEDs also are used in a large number of other applications, from projectors and flashlights to car tail- and headlights and general illumination. Solid-state white-light sources can be realized either by mixing different LEDs emitting red, green or blue light, or by using a phosphor material to convert monochromatic light from a blue or UV LED to broad-spectrum white light.

With the increase in LED production, manufacturers are looking for new process developments to optimize scribe width, speed and production throughput. New advances in laser liftoff (LLO) and laser wafer scribing for LEDs provide manufacturers with cost-effective industrial tools that are ready to meet increased demands.

High-brightness LEDs with vertical structure

Typically, blue/green LEDs are composed of a GaN film a few microns thick, grown epitaxially on a sapphire substrate. Some of the major costs of LED fabri-

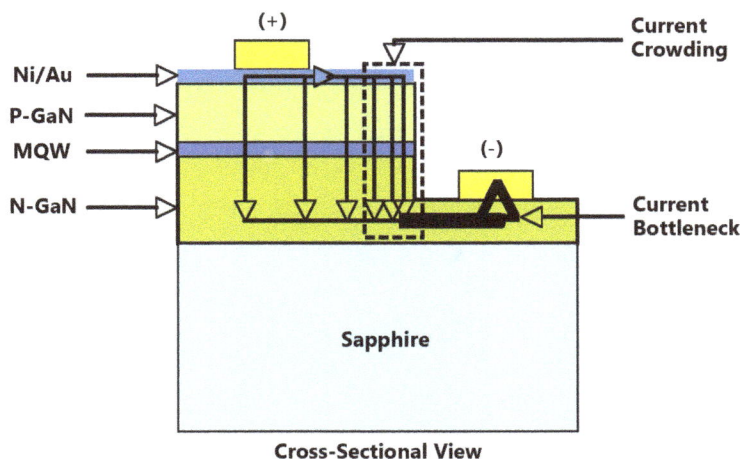

Figure 1. This diagram shows the traditional horizontal configuration for a blue LED. MQW = multiple quantum well.

cation are the sapphire substrate itself and the scribe-and-break processes. For the traditional LED configuration, the sapphire is not removed, so both cathode and anode are installed on the same side of the GaN epitaxial (epi) layer (Figure 1).

There are several drawbacks to this configuration. For high-brightness LEDs, disadvantages include a high current density inside the material, current crowd-

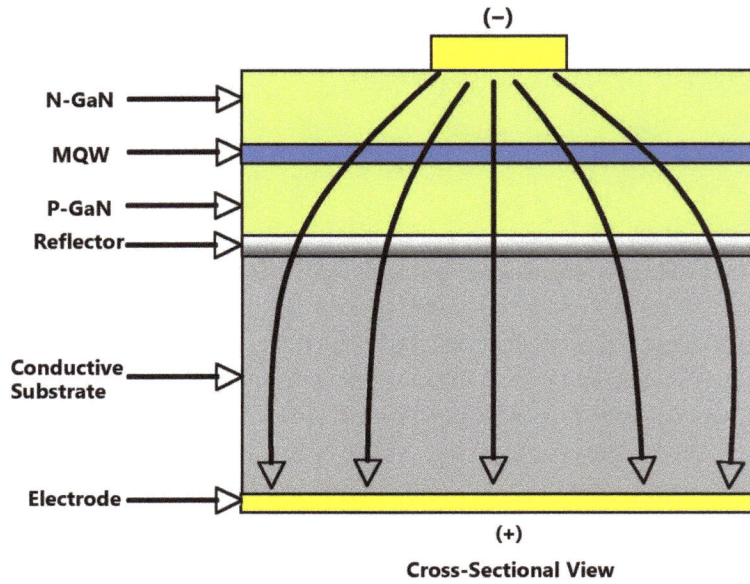

Figure 2. A model of a vertical configuration for a blue LED is shown.

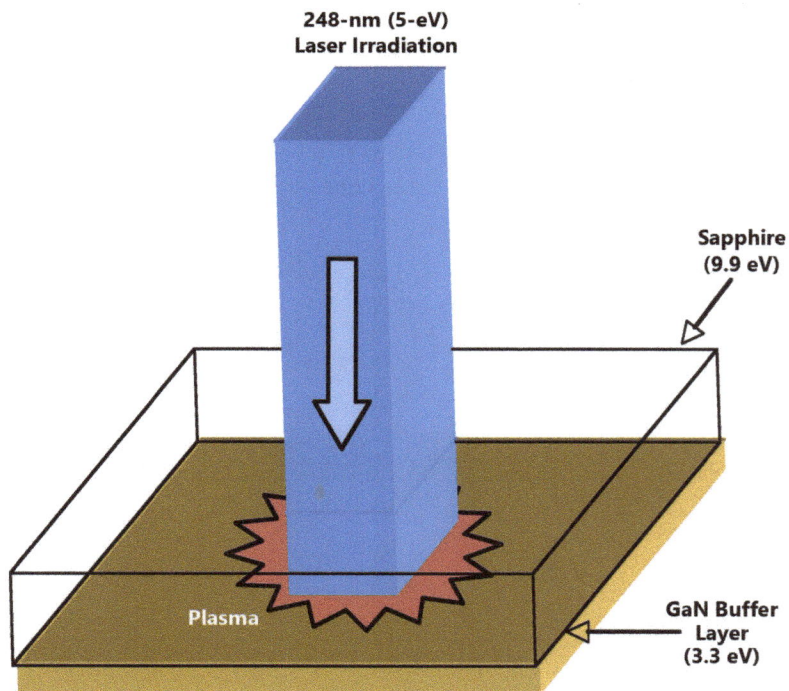

Figure 3. This is a schematic representation of laser liftoff at 248 nm.

Figure 4. This diagram illustrates single-pulse laser liftoff of GaN from sapphire at 248 nm (one pulse covers nine die).

ing, reduced reliability and shorter lifetimes. Also, there is significant light loss through the sapphire.

By using an LLO process, LED designers can create a vertical LED, which overcomes many of the limitations of the traditional horizontal configuration. Vertical configuration provides the possibility of pumping an LED with more current, eliminating the undesired current crowding and bottleneck inside the device and significantly increasing the maximum light output and efficiency of the LED (Figure 2).

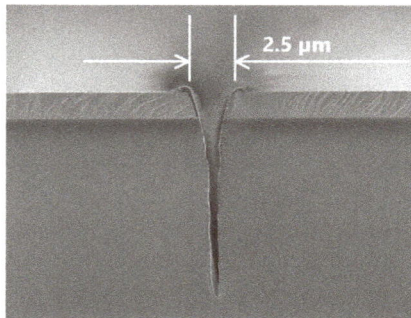

Figure 5. Shown here is the kerf width in a GaN-on-sapphire wafer.

The vertical LED structure requires removing the sapphire before attaching the electrical contacts. Excimer lasers have proved to be valuable tools for separating the sapphire and GaN thin film. LED laser liftoff dramatically reduces the time and cost of the LED fabrication process, enabling the manufacturer to grow GaN LED film devices on the sapphire wafer and to transfer the thin-film device to a heat sink electrical interconnect. The process allows for creation of freestanding GaN films and integration of GaN LEDs onto virtually any carrier substrate.

Laser liftoff principle

The basic concept behind UV LLO is to use the different absorptions of UV laser light in the epi material and the sapphire. With a high (9.9-eV) bandgap energy, sapphire is transparent to 248-nm KrF excimer laser radiation (5 eV), whereas GaN (approximately 3.3-eV bandgap) strongly absorbs the 248-nm laser light. As shown in Figure 3, the laser light travels through the sapphire and couples with

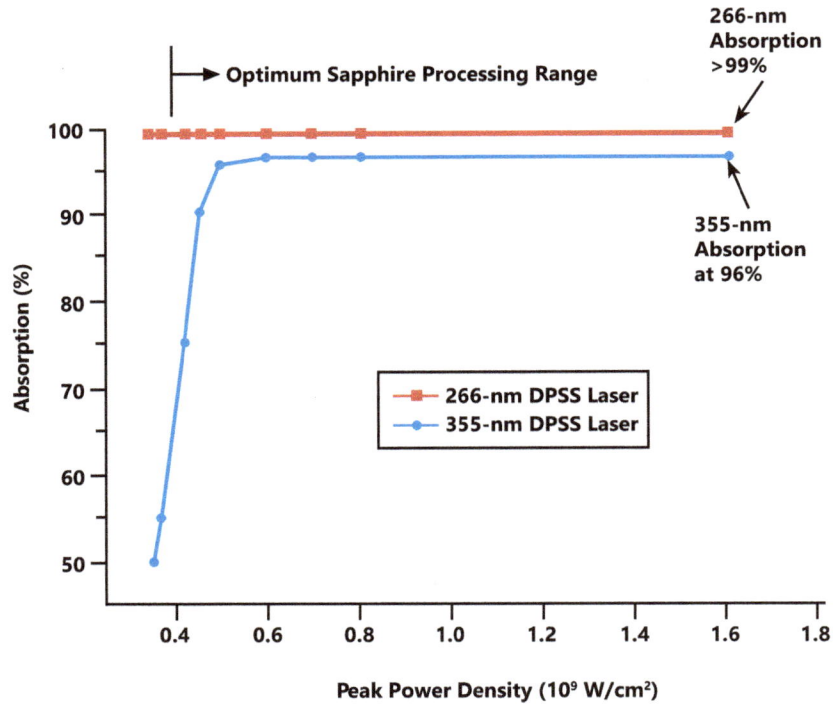

Figure 6. This graph represents UV absorption in LED sapphire.

Optimum Sapphire Processing Range

266-nm Absorption >99%

355-nm Absorption at 96%

266-nm DPSS Laser
355-nm DPSS Laser

Absorption (%)

Peak Power Density (10⁹ W/cm²)

the GaN, causing ablation at the GaN-sapphire interface. This creates a localized explosive shockwave and debonds the GaN from the sapphire at that location. The same principle applies to AlN on sapphire when using 193-nm ArF excimer laser radiation. Aluminum nitride with a bandgap of 6.3 eV can absorb the 6.4-eV ArF radiation, but AlN is still transparent to sapphire with a 9.9-eV bandgap.

To achieve successful liftoff, both beam homogeneity and wafer preparation are important. At J.P. Sercel Associates (JPSA) Inc., innovative and patented beam homogenization techniques with excimer lasers create a flattop beam on the wafer with uniform energy density distribution across an area as large as 5 × 5 mm.

Correct wafer preparation is crucial for successful LLO. It minimizes the residual stress from the high-temperature epi layer growth on sapphire, and it ensures adequate bonding between the epi layer and the carrier substrate to avoid fractures along the epi during liftoff. Figure 4 shows a typical liftoff result.

Using LLO systems leads to high-speed, high-yield production at ambient temperature. Well-designed systems allow exposure of multiple die simultaneously with a single shot and permit accurate placement of each shot across the wafer using a novel "fire on the fly" technique.

Blue LED wafer scribing

There are traditional manufacturers who continue to supply horizontal-structure blue LEDs, and laser scribing is ideal for processing this wafer configuration. The extreme hardness of the sapphire causes significant problems for both saw dicing and diamond scribing, including low die yield, low throughput and high operating costs.

The use of UV diode-pumped solid-state (DPSS) lasers has proved to dramatically increase die yields and wafer throughput as compared to traditional diamond scribing methods, without appreciable loss of brightness in LED wafers. The short wavelength enhances optical absorption at both the GaN and sapphire layers, lowering the irradiance required for ablation while simultaneously allowing for reduced cut width.

Scribe width, speeds and production throughput are essential to keeping manufacturing costs low and wafer yields high. JPSA has developed a patented beam delivery system that allows for a very narrow kerf of 2.5 µm wide (Figure 5) and offers proprietary surface protection that minimizes debris. Moving the wafer under a tightly focused laser beam produces an extremely narrow V-shape cut; starting at the epi side and extending into the sapphire, it is typically 20 to 30 µm deep. After laser scribing, V-shape laser cuts act as stress concentrators for the process of breaking with standard cleaving equipment.

The narrower kerf width generated by 266-nm front-side scribing increases the number of usable die produced per wafer, potentially boosting the entire fab operation.

An easy comparison involves a typical 2-in. blue LED wafer on sapphire, with 250×250-µm devices. Using traditional diamond scribing with typical 50-µm streets (300-µm die pitch), there will be approximately 22,500 die on the wafer. The typical breaking yield is 90 percent for traditional diamond scribing and results in 20,250 usable die per wafer.

By using UV laser scribing, the street width can be reduced to 20 µm (a 270-µm pitch), increasing the number of die on the wafer to approximately 27,800 (a 23 percent increase). With increased breaking yields, the method produces about 27,500 usable die — a 35 percent total increase in usable die per wafer.

Since 1996, JPSA has been using 266-nm DPSS lasers to scribe blue LED sapphire wafers from the GaN front side at speeds of 150 mm/s, leading to a throughput of about 15 wafers per hour (for standard 2-in.-diameter wafers with a die size of 350×350 µm). With its high throughput and minimal impact on LED performance, the process is tolerant of wafer warp and bow, delivering much faster scribing speeds than traditional mechanical methods.

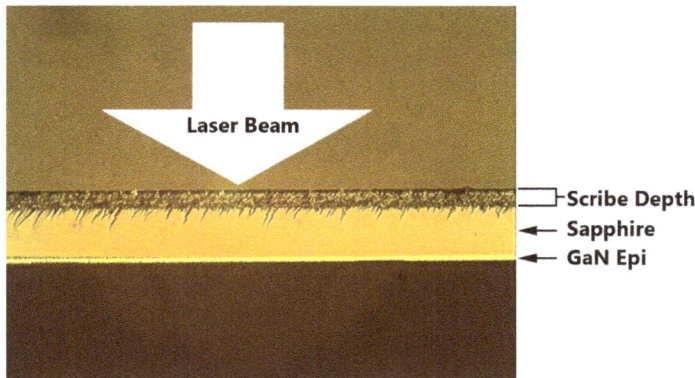

Figure 7. Shown here is a cross section of a GaN wafer scribed from the sapphire (back) side by a 355-nm diode-pumped solid-state laser.

Scribing silicon carbide

In addition to sapphire, silicon carbide can be used as an epitaxial growth substrate for thin blue LEDs. Ultraviolet DPSS lasers at 266 and 355 nm (4.6 and 3.5 eV, respectively) excel at scribing silicon carbide, which has a large bandgap of around 2.8 eV. Because of the high photon energy, enhanced coupling is achieved, allowing for high-speed scribing and easy breaking. Thick III-nitrides such as GaN and AlN also can be scribed using UV DPSS lasers. While the scribing speed for 200- to 400-µm-thick GaN or AlN is significantly reduced compared with the speed for thin epi films on sapphire or silicon carbide, the cut quality is excellent and allows for a clean breaking step.

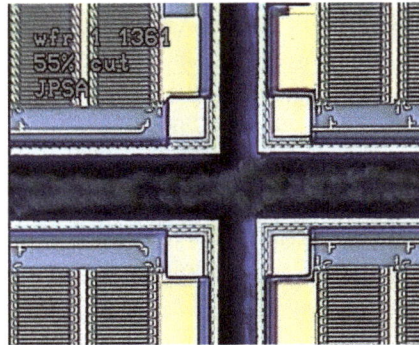

Figure 8. Clean, well-defined edges are highlighted in this scribed and expanded GaAs wafer.

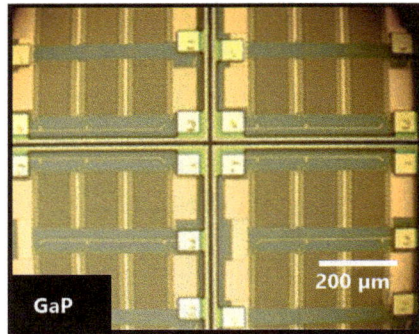

Figure 9. Pictured is GaP scribing at 300 mm/s for a 30-m cut, which is deep enough to break wafers up to about 250 µm thick.

For vertical high-power LEDs, LLO detaches the sapphire while the epi film remains bonded, typically to a high-conductivity carrier substrate such as copper, copper-tungsten, molybdenum or silicon. For wafer-based silicon, cutting depths of 100, 150 and 200 µm can be achieved at 300, 150 and 100 mm/s, respectively. Beam delivery techniques allow these scribing speeds/depths while requiring only limited laser power and, consequently, minimizing thermal affectation. Scribing of metal-based wafers is challenging because of high thermal conduction that typically leads to a weld back effect. In addition, a full cut often is required for separation because the material is extremely ductile. At JPSA, we have developed advanced techniques that allow for successful scribing of these substrates up to 200 µm thick and that could prove extremely important for the high-brightness LED industry.

Dual scribing capability

Back-side scribing by 355-nm DPSS lasers allows scribing from the sapphire side of the LED. Wafer alignment can be performed from the front or back side using multiple inspection cameras, which is important if the sapphire has a metallic reflective layer. Also, the epi is not directly exposed to laser radiation, which may minimize light loss. When compared to a 266-nm laser, the longer 355-nm wavelength leads to reduced absorption in the sapphire (Figure 6). As a consequence, higher power typically is required, which leads to wider kerfs and streets. Additionally, back-side scribing is applicable only for sapphire wafers <150 µm thick, while front-side scribing allows for thicker wafers that may be lapped down to the final thickness required for breaking after processing.

With continued research and development in back-side scribing, such as new laser absorption enhancement techniques, JPSA has achieved high-throughput back-side scribing at speeds of up to 150 mm/s with no debris or damage to the epi layer (Figure 7).

Wafer scribing III-V semiconductors

An alternative method for separating brittle compound semiconductor wafer materials in GaAs, InP and GaP wafers is scribing with UV DPSS lasers that rapidly process wafers with kerfs around 3 µm with no edge chipping in any of the III-V materials, making straight, accurate and clean cuts (Figure 8). Typically, wafers up to 250 µm thick are scribed at 300 mm/s, allowing for easy breaking (Figure 9). The III-IV wafers are expensive, so wafer real estate is valuable. The tighter, narrower and cleaner cuts achieved using UV lasers provide a better die count per wafer as well as higher yields, due to fewer damaged die than with conventional saw scribing methods.

Outlook

LED technology is advancing rapidly as it strives for higher efficiency and lower manufacturing costs. This "green" technology undoubtedly has a bright future; however, it also faces considerable challenges.

The current explosion of worldwide demand for LEDs requires the development of new laser processes and technologies to produce even higher quality, yield and throughput. In addition to the ongoing development of laser systems, new machining techniques and applications, improved beam delivery and optical systems, and an enhanced knowledge of the interactions between laser beams and materials will be required to sustain the progress of this green revolution.

Equipment engineers are challenged to build flexible tools. Options, such as the capability for automated cassette loading and unloading, edge detection and auto-focus systems, define the state-of-the-art laser scribing solution. Companies such as JPSA continue to explore the forefront of laser technology to meet the demands of the LED manufacturing market.

Meet the authors

Jeffrey P. Sercel is chairman and chief technology officer of JPSA in Manchester, N.H. Marco Mendes is director of JPSA's applications lab.

Micromachining

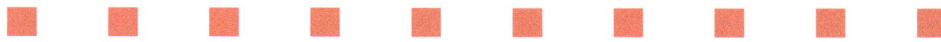

Femtosecond Lasers: More Micromachining Apps

From medical implants to smartphones to automotive uses, femtosecond lasers are a reliable tool for industry

BY DANIEL ACHENBACH AND VICTOR MATYLITSKY,
SPECTRA-PHYSICS INC.

In the more than two decades since a team of researchers from the Center for Ultrafast Optical Science at the University of Michigan first demonstrated the use of a femtosecond laser for micromachining[1], this technology has clearly made its evolutionary step from scientific equipment to a reliable tool for industrial manufacturers. With their ability to process any material with a minimal amount of heat-affected zones (HAZ), femtosecond lasers are being considered for a growing list of micromachining applications.

The short pulse duration of femtosecond lasers enables materials processing with cold ablation, and the optional second harmonic allows smaller features or higher ablation rates. Such short pulse durations, along with higher energies and lower costs, are helping femtosecond lasers produce the next generation of medical implants, make smartphone glass covers more durable and improve the fuel efficiency of automobiles through the drilling of gasoline injector nozzles.

Stent cutting

The manufacturing of stents requires machining of small feature sizes down to a few micrometers with an edge quality that does not show burring, melting and recast. In addition, heat deposition in the material results in a HAZ bordering the cut edges. Within the HAZ, material properties or composition are altered.

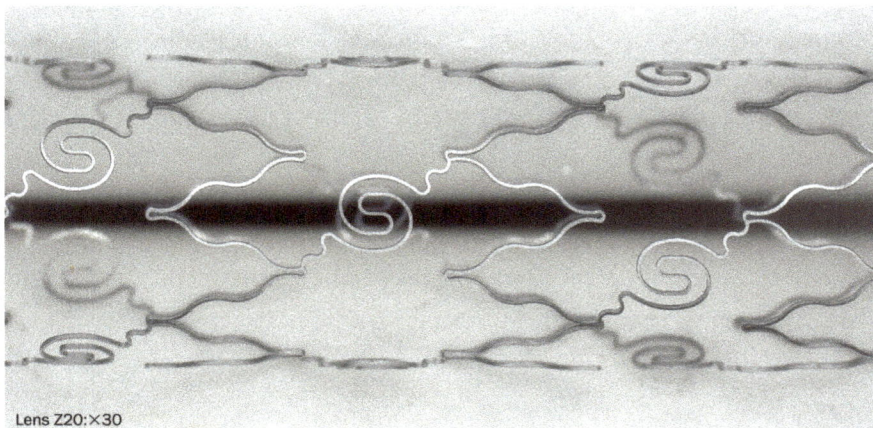

Figure 1. Metal alloy stent made of nitinol machined by femtosecond laser. Courtesy of Spectra-Physics Inc.

Consequently, cleaning, deburring, etching and final polishing are routinely employed to bring the stent's surface properties to the level and consistency required for implantable devices. Some of these post-processing steps could be avoided by fabricating stents with femtosecond lasers (Figure 1).

A new generation of implants are biodegradable stents made of bioabsorbable polymer tube materials (Figure 2). These materials are very sensitive to induced heat, as the melting temperature is usually below 200 °C. Because of the low melting point of the bioabsorbable polymers, application of a high average power femtosecond laser for machining of bioabsorbable materials would lead to the formation of HAZ along of the cutting edge.

On the other hand, the cutting speed can be increased by using a high pulse energy femtosecond laser without affecting the quality of the cut. An additional boost in throughput can be obtained by using laser pulses at shorter wavelengths. Upon application of these laser pulses, laser energy can be used more efficiently for ablation of transparent materials. For example, by applying laser pulses at 520 nm, the speed of athermal cutting is at least a factor of two higher in comparison with the results obtained by using laser pulses at 1040 nm (Figure 3).

By choosing the optimal process parameters — pulse energy, repetition rate, pulse overlap and wavelength — and adding assist gas, the cutting speed of up to 50 mm/sec could be demonstrated in application tests.

An important factor for good process quality is the pulse duration. The series of application tests have shown that HAZ is already increased by moving from 400 fs to 800 fs. A further increase to 10 ps results in low cutting speed (Figure 3) and large melting zones. Machining of medical devices is a good example for a process that requires an ultrafast laser with short pulse duration (less than 400 fs) and high pulse energy.

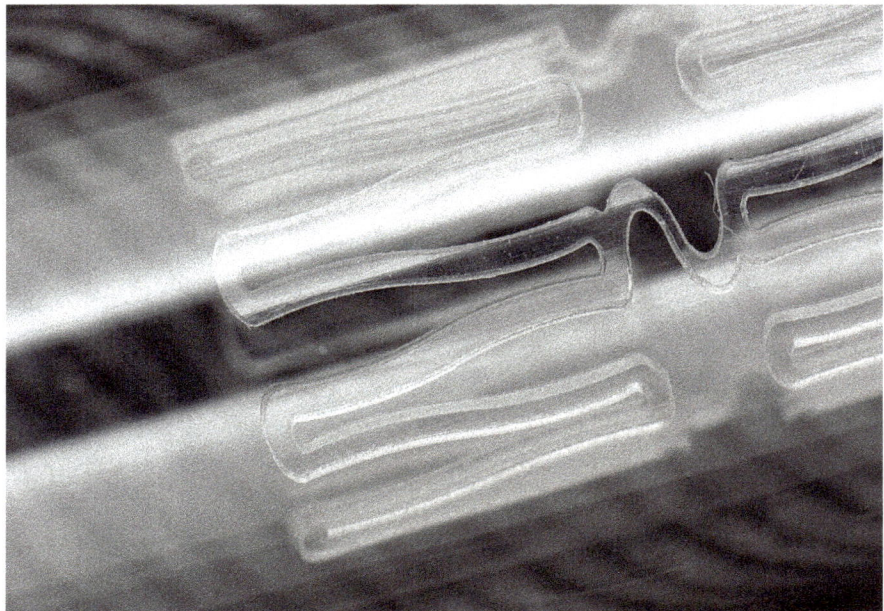

Figure 2. Polymer stent made of PLLA (polylacitide) machined with femtosecond laser. Courtesy of Spectra-Physics Inc.

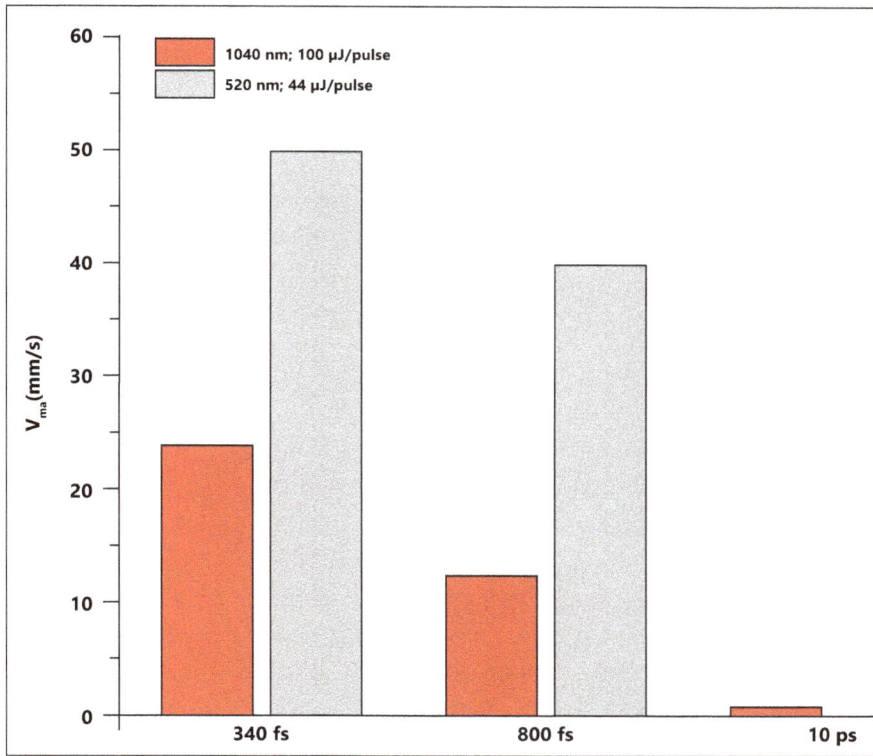

Figure 3. Cutting speed for PLLA processed with Spirit laser depending on pulse duration and wavelength. Courtesy of Spectra-Physics Inc.

Cutting of transparent, brittle materials

Recently, femtosecond lasers have become popular tools for machining transparent, brittle materials, as the ability of cold ablation promises process results with minimum amount of chipping and micro-cracks. Cutting of chemically strengthened and nonstrengthened glass for cover glasses for mobile electronic devices is a primary market driver. Application of femtosecond lasers can result in high cut quality — defined by the average roughness of the cross-section — directly after the cutting process to save post-processing steps that have been necessary after a mechanical cut. The very high cutting quality also leads to extremely high bending strength.

Although glass is transparent for near-infrared and visible wavelengths, femtosecond pulses can interact with this matter, if the laser beam is focused tightly enough. The high peak intensity of femtosecond lasers enables nonlinear absorption inside of transparent materials. Microdamage can be induced inside the bulk material, causing crack propagation along the cutting pass.

Excellent bending strength with minimal chipping occurs with Spectra-Physics Inc.'s patent pending ClearShape process that uses a femtosecond Spirit laser with a moderate output power of 4 W with 40 µJ at 1040 nm (Figures 4 and 5).

To achieve the controllable cleaving along the curvilinear cut, a multiscan cut is often needed. The larger number of scan passes along the cutting path decreases

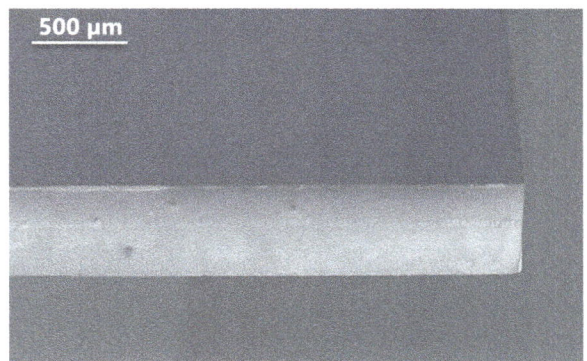

Figure 4. Example of straight line cut in chemically strengthened Gorilla Glass from Corning. Courtesy of Spectra-Physics Inc.

the effective cutting speed. The availability of femtosecond lasers with more than 100-µJ pulse energy and special multifoci optics now enable simultaneous modification of four layers, resulting in four-times-faster cutting speeds.

Nozzle drilling

Femtosecond lasers' high precision and excellent process quality is ideal for drilling gasoline injector nozzles. To achieve a high-efficiency engine, the method of injecting gasoline or diesel into the combustion chamber is very important. The more homogenously the fuel is sprayed into the chamber, the more efficiently the fuel can be used to run the engine. By optimizing the spray pattern inside the chamber, the engine's efficiency can be increased, resulting in reduced fuel consumption.

The spray pattern depends on the injection pressure, but also on the geometry and sidewall quality of the nozzle holes. Hence, these holes must have very smooth walls post-drilling. Historically, these tiny and high aspect ratio holes with 150- to 250-µm diameters have been drilled by electron discharge machining (EDM). Until a few years ago, femtosecond lasers could not yet provide the industrial reliability and the essential pricing to be competitive with the traditional EDM method. However femtosecond lasers have now reached levels of reliability and pricing so that they can be dependably used in automotive production.

The process of drilling small, high aspect ratio holes with excellent surface quality requires ultrafast lasers with high energy pulses of 80 µJ or more at ultrashort pulse durations. The latest experiments have shown a large quality increase by reducing the pulse duration from approximately 10 ps down to less than 400 fs (Figure 6). For the drilling of very narrow holes, higher pulse energies at lower repetition rates is more beneficial than higher output powers and higher repetition rates. Too many pulses per second or too much average power would result in HAZ that inhibits a smooth machined surface. Therefore, an average output power of 10 W is suitable for the gasoline nozzle drilling process. For drilling holes with high aspect ratio, a shorter wavelength, such as the second harmonic of a ytterbium-based laser at around 520 nm, is beneficial. The advantages are a smaller focus spot size and a larger Rayleigh length. This enables higher aspect ratios in the drilling process.

Beside the sidewall roughness, the edge quality and geometry of the nozzle hole's exit to the combustion chamber is even more important. The fuel spray pattern strongly depends on the geometrical form of the transition area between the nozzle holes and the chamber. A 90° edge will not provide the most homogenous distribution of the fuel. To optimize the injection process, the drilling procedure needs additional degrees of freedom to generate an edge geometry different from 90°. For the laser drilling process it is necessary to tilt the beam

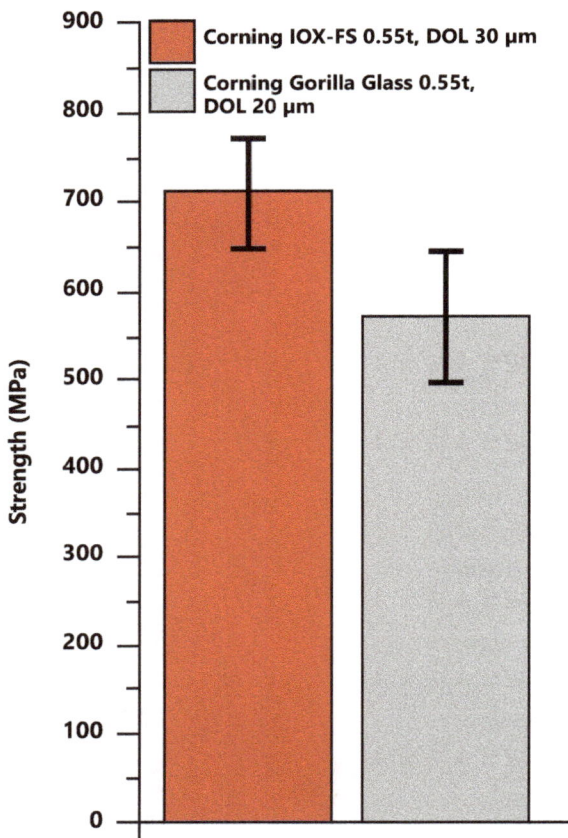

Figure 5. Results of four-point bending tests for chemically strengthened glasses from Corning (right) obtained using the ClearShape process with a Spirit femtosecond laser. Courtesy of Spectra-Physics Inc.

Injection nozzles produced around the world in 24/7 production

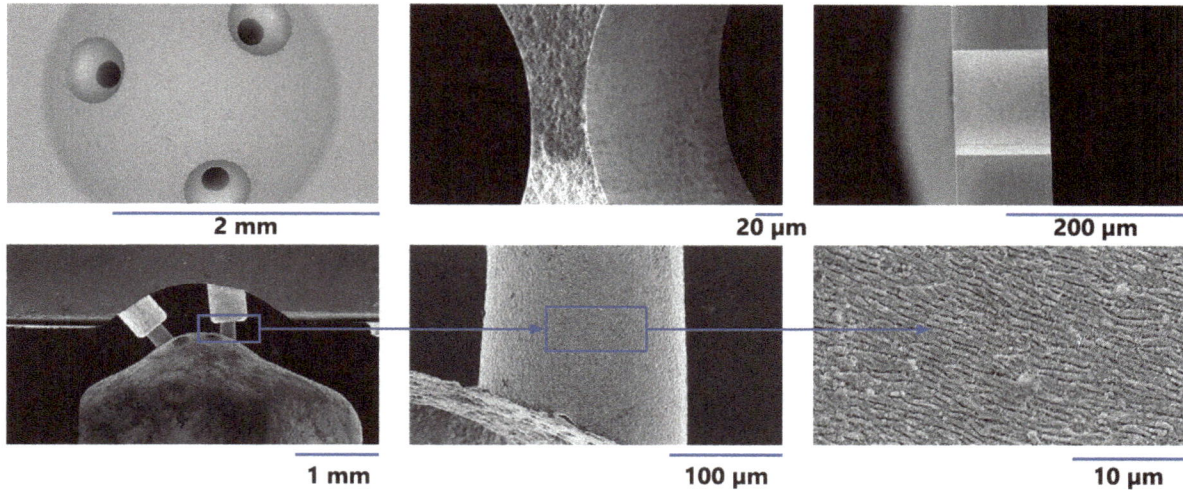

2 mm	20 µm	200 µm
1 mm	100 µm	10 µm

Typical material thickness	0.02-1.5 mm	**Hole aspect ratio**	> 12:1	**Hole circularity**	> 95%
Typical hole processing time	1.5 s	**Diameter repeatability**	< 0.25% variation	**Hole position accuracy**	< 1 µm
Hole diameters	25-700 µm	**Hole diameter accuracy**	< 1 µm	**Wall surface quality Ra**	< 0.1 µm
Negative taper full angle	> 16°				

by a few degrees to the perpendicular of the nozzle surface. This will also have the advantage that the workpiece does not need to be moved during the whole drilling process.

For this advanced drilling application, a special galvo scanner with precession technology is necessary. Successful results were demonstrated with ARGES GmbH's Precession Elephant scanners (Figure 7) that are already used in automotive production at most injection nozzle manufacturers. Because of the excellent quality and repeatability, the traditional EDM method is mostly replaced by this laser scanning technology. Importantly, the controllable hole taper helped increase the fuel efficiency by up to 20 percent in modern engines.

The implementation of the laser drilling process for diesel nozzles is the next development step. Diesel nozzles have a higher wall thickness, as the injection pressure in diesel engines is much higher than in gasoline engines. Consequently, holes with even higher aspect ratios need to be drilled. For this application, ultrafast lasers with pulse energies >40 µJ at wavelengths in the visible range (e.g., 510 to 530 nm) will be necessary to substitute for conventional EDM methods.

Figure 6. Samples of injection nozzles processed with the ARGES GmbH Precession Elephant scanner and femtosecond laser. Courtesy of ARGES GmbH.

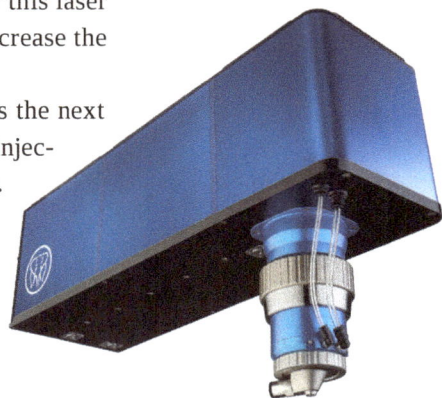

Figure 7. ARGES' Precession Elephant scanner. Courtesy of ARGES GmbH.

Femto's future

In the coming years, femtosecond lasers will continue to improve in cost-performance, resulting in their application in new market segments, as lasers become even more competitive to mechanical machining methods. Femtosecond lasers will provide higher average power and pulse energies for higher throughput, whereas the pulse duration has already reached its sweet spot for most applications.

Meet the authors

Daniel Achenbach is the product manager for femtosecond amplifiers at Spectra-Physics Inc. in Rankweil, Austria; Victor Matylitsky is the business development manager for ultrafast laser applications at Spectra-Physics in Rankweil, Austria.

Acknowledgments

The authors would like to thank our application expert, Frank Hendricks, for his research work on machining of polymer stents and brittle materials. We also thank our partners at ARGES GmbH for collaborating with us in the drilling applications.

Reference

1. R.R. Gattass (2008). Femtosecond laser micromachining in transparent materials. *Nat Photonics*, Vol. 2, pp. 219-225.

Ultrashort Pulse Laser Micromachining Surpasses Previous Limitations

With the power of ultrashort pulse laser systems on the rise, achieving dynamically and synchronously adaptable pulse repetition rates in the MHz range is the key to higher throughput.

BY FLORIAN HARTH, THOMAS HERRMANN, BERNHARD HENRICH AND JOHANNES A. L'HUILLIER PHOTONIK-ZENTRUM KAISERSLAUTERN EV AND RESEARCH CENTER OPTIMAS

To fulfill current and future customer needs in the ultrashort pulse (USP) laser micromachining market, faster processing time is critical. In only the last few years the average power of these lasers has risen continuously, reaching an average power of 1 kW or more. The higher output power allows for an increased pulse repetition rate (PRR), while maintaining the necessary pulse energy for efficient material ablation. This means more ablation per second, strongly increasing throughput.

However, with a highly increased PRR comes a new challenge: It's increasingly difficult to avoid the accumulation of too many pulses on one spot of the workpiece. The ultrahigh PRR of modern laser systems simply exceeds the deflection capability of galvanometer-based scanning systems. Resonant and other scanner

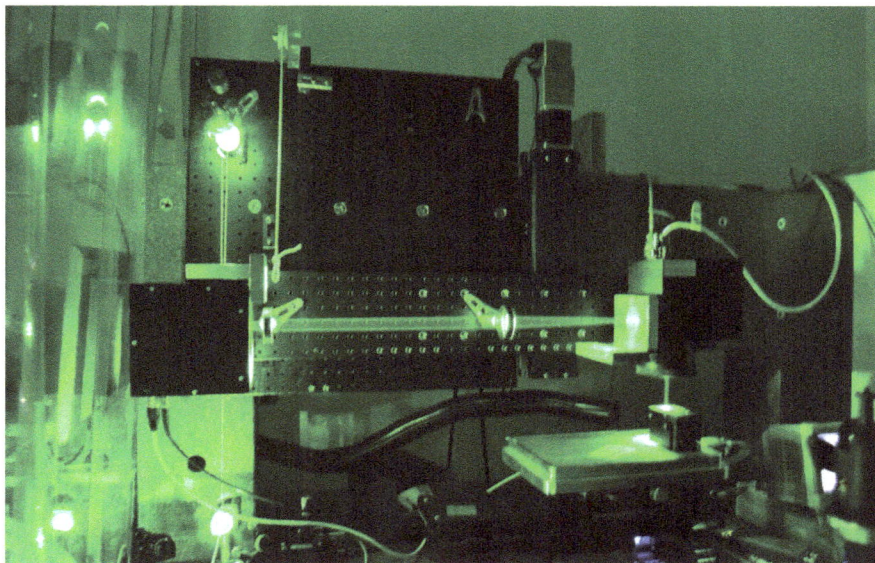

Initial lab setup for ultrafast micromachining with a resonant scanner and the new laser system. Images courtesy of Photonik-Zentrum Kaiserslautern eV.

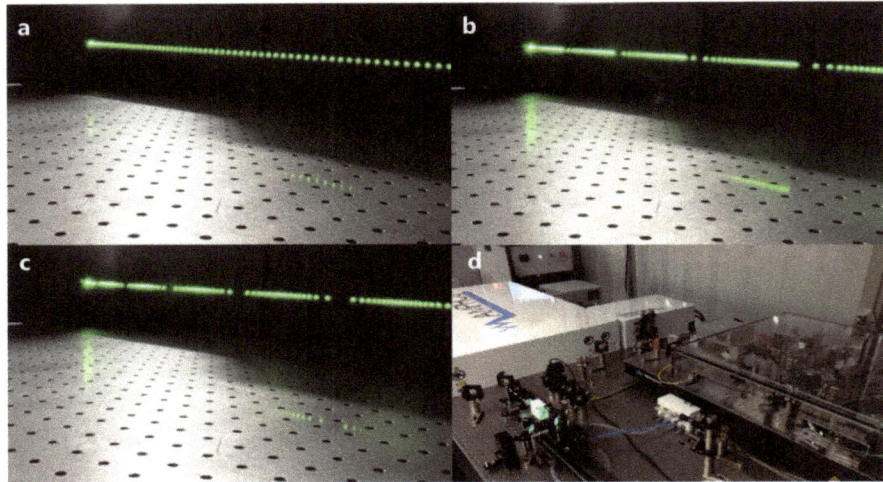

Figure 1. Schematic picture of a pulse train with (a) a constant pulse repetition rate (PRR), (b) a linear sweep of the PRR, (c) a periodic modulation of the PRR and (d) a picture of the lab setup of the prototype.

technologies, reach extremely high scan speeds of more than 1000 m/s, but they suffer from a sinusoidal varying scan speed, leading to an inconsistent spot distance on the workpiece.

To overcome this drawback, the nonlinearity of the resonant scanners can be compensated by a dynamically and synchronously adaptable PRR. In this way a uniform spot distance across the whole scanning range becomes possible. Modern laser systems with dynamically adaptable pulse repetition rates, combined with ultrafast resonant scanners, are great candidates for breaking through the current limitations of high-speed micromachining.

USP laser systems with dynamic PRRs have shown great potential in other applications, like speeding up the processing of narrow curves[1], or homogenizing the ablation with a laser turning machine[2]. In that case, a variable PRR in the kHz range is achieved by picking pulses out of a pulse train with a given base PRR. An important consideration for future applications is whether it is possible to scale this approach to the MHz range.

The following addresses the requirements, possible approaches and a working setup for next-generation USP lasers, capable of adapting the pulse repetition rate in the MHz range with a very fine resolution. The requirements will be specified for the resonant scanner, since these are the highest among the possible applications.

Dynamically variable pulse repetition rate

The laser deflection speed in the focal plane of a resonant scanner-based micromachining setup can be described by a sinusoidal function. Since the spot distance is proportional to the scanning speed ($\Delta s = v/\text{PRR}$), a repetition rate, which changes proportionally to the deflection speed, leads to a fixed spot distance across the whole scanning range. Moreover, since a resonant scanner typically oscillates in the 10-kHz range, the laser's PRR has to be changed more than 10,000 times per second. Due to the very high velocity of the deflected beam of about 1000 m/s, a high PRR in the MHz range is required to reach an adequate pulse overlap. In order to maintain high accuracy and a precise positioning on the workpiece, the PRR ideally has to be tuned continuously in this range.

In a mode-locked-based USP laser system, the PRR is given by the cavity round-trip time of the laser oscillator. Therefore, the variation of the PRR of these USP lasers is done by pulse picking with the help of fast electro-optical or acousto-optical modulators, suitable for high-power operation. The base PRR can be divided by integers and is reduced to an effective PRR. This is shown in Figure 2 for a base PRR of 10 MHz, 100 MHz, 1000 MHz and a true arbitrary pulse-on-demand operation.

A typical USP laser, operating at a fundamental PRR of 100 MHz, offers very few effective PRRs in the MHz range. If the application requires a PRR variation between 10 MHz and 5 MHz, only 11 individual repetition rates are accessible. This is not enough to ensure high positioning accuracy after the resonant scanner. Only a much higher base PRR — or even an arbitrary pulse-on-demand technology — can fulfill this requirement (Figure 3).

The accuracy of the spots for a base PRR of 100 MHz is only $\Delta s = \pm 4.1$ µm, which is too poor for high precision surface structuring with focus diameters in the 10-µm range. A base PRR of 1 GHz allows for a much higher accuracy of $\Delta s = \pm 0.4$ µm. This accuracy is competitive with current technologies and sufficient for USP laser micromachining applications.

This calculation also holds true for speeding up the processing of narrow curves or homogenizing the ablation with laser turning machine applications. Processing narrow curves with variable PRRs in the 100-kHz range already ensures high accuracy, since the granularity is fine enough (compare to Figure 2). A true scaling to the MHz range, however, is only achieved if the base PRR is strongly increased or an arbitrary pulse-on-demand setup is used. Increasing the throughput of these applications by an order of magnitude leads to a reduction of a 10-second production process to just one second.

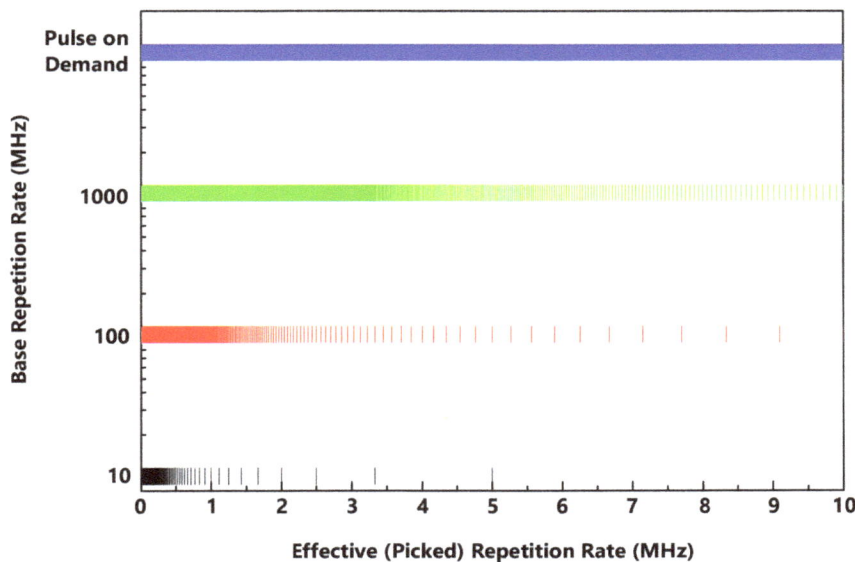

Figure 2. Achievable repetition rates with different seed sources by pulse picking. A higher base PRR enables more possible effective PRRs in the upper MHz range.

Evaluating semiconductor lasers, ultrafast pulse picking

Using a high-base PRR with ultrafast pulse picking or an arbitrary pulse on-demand technique are the best approaches. A pulse repetition rate in the GHz range requires a short cavity length of a few mm, making semiconductor lasers a feasible option. However, a key challenge is the amplification of these pulses, with energies in the range of a few pJ, to the required µJ pulse energies. In general, a regenerative amplifier would provide the necessary amplification, but the high and dynamical variable PRR would get lost. So the most feasible way is a combination of a semiconductor oscillator and a linear amplifier chain, consisting, for example, of a fiber pre-amplifier and a solid-state power amplifier. A combination of the best concepts derived from different state-of-the-art technologies will do the job.

USP semiconductor lasers with PRRs between 1 to 10 GHz and ultrafast pulse picking were developed years ago[3] and since then the technology has matured. A mode-locked semiconductor laser oscillator with a base PRR of 4.3 GHz, for instance, was developed by FBH Berlin[4]. Mode locking was achieved in a monolithic Fabry-Perot diode laser resonator. Pulse picking was done by ultrafast pumping of a subsequent waveguide preamplifier beyond the transparency level and back. Gate widths of 200 ps were achieved, short enough for picking single pulses out of the 4.3-GHz pulse train. The setup delivers ultrashort pulses in the 10-ps range with adjustable PRR.

If there is a demand for finer granularity, operators can leave the mode-locking technique and use gain-switching instead. Gain-switching delivers arbitrary pulses on demand, so a discrete base PRR is no longer a concern. The accuracy, then, is only limited by the temporal jitter between the trigger and the emission time of the corresponding pulse, which typically is extremely low. Looking back to the resonant scanner application, the spot distance Δs would be absolutely constant across the whole working range (Figure 3).

Figure 3. Variation of the repetition rate as the resonant scanner scans the workpiece. As the scan-speed decreases at the turning points, a lower PRR is necessary to maintain the spot distance Δs. Only discrete values are possible for a low base PRR (a). Discrete repetition rates lead to large deviations from the desired spot distance Δs = v/PRR, where v is the continuously varying scan-speed (b).

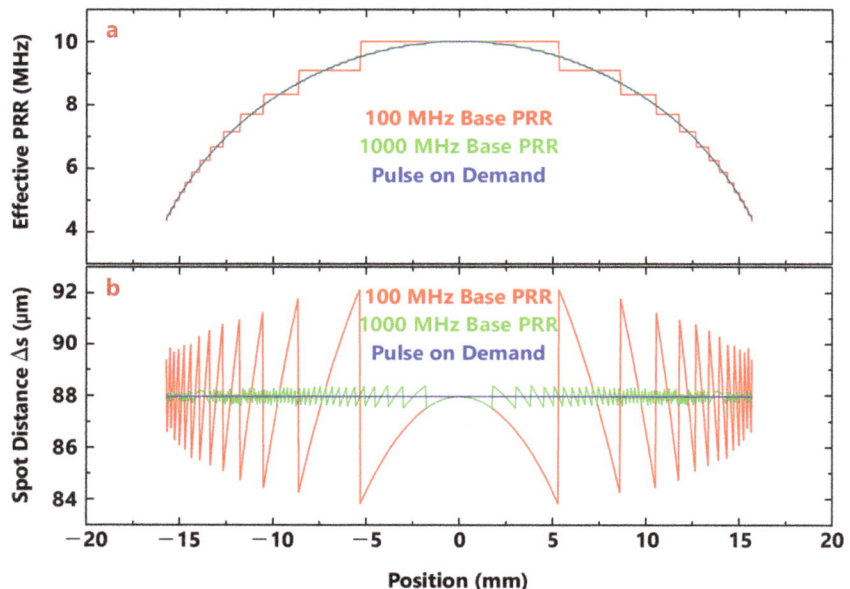

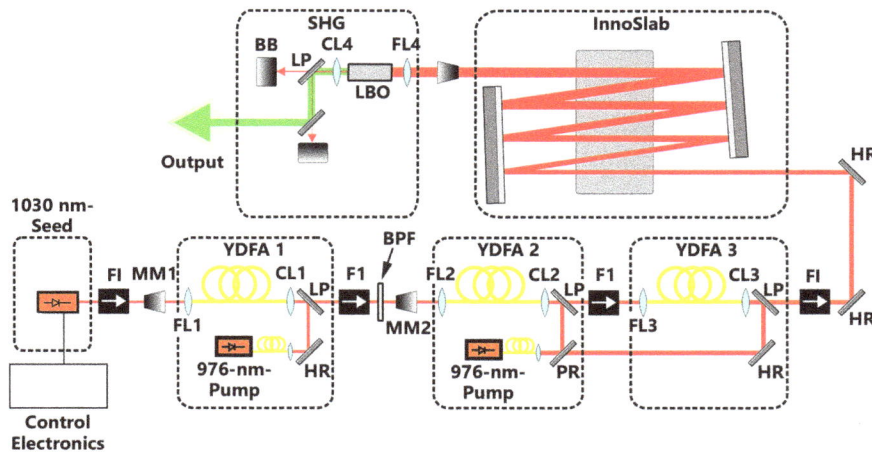

Figure 4. Schematic setup of a highly dynamic laser system. The fiber pre-amplifier was optimized to provide a large small-signal gain combined with minimized amplified spontaneous emission. The output power was 200 W in the IR or 130 W after an optional SHG stage.

The drawback of gain-switching diodes is pulse length. The shortest pulses, directly obtained by gain-switching, are in the range of approximately 40 ps. Many applications would be adversely impacted by the longer pulse duration, in terms of ablation efficiency and thermal impact. Nevertheless, the authors showed that even processing of transparent media is possible with these pulses[5].

Figure 4 shows a schematic of an actual setup, based on gain-switching, that was recently demonstrated[6]. The amplifier chain consists of a fiber preamplifier, and an InnoSlab power booster. An optional SHG-stage can convert the 1030-nm radiation to the green spectral range, depending on the application.

Varying pulse energies

A crucial issue for most applications is maintaining a constant pulse energy while changing the pulse repetition rate. Since the PRR is altered in front of the

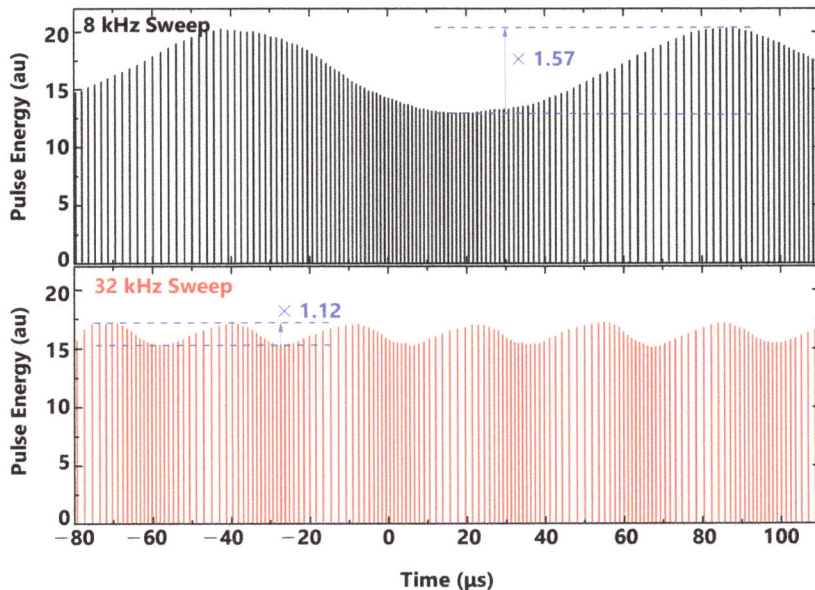

Figure 5. Temporal evolution of the pulse energy for PRRs between 5 MHz and 10 MHz at different sweep-frequencies. For visibility reasons only every 10th pulse is plotted.

amplifier chain, a closer look at the pulse energy dynamics is required. In general, a lower PRR would result in higher pulse energy for a constant average output power. Varying pulse energies during large-scale processing, however, would have a strong impact on the quality of machined parts.

The very high processing speed helps in addressing this problem. On the one hand, typical laser active media, commonly used in the amplifier chain, exhibit a long upper state lifetime in the ms range. Conversely, the laser's PRR changes more than 10,000 times per second, which means that the repetition rate changes so fast that the amplifier cannot adapt to the new situation. An average, and nearly constant, pulse energy, independent of the PRR, is emitted.

In practice, the relaxation time of the overall amplifier chain depends on a number of parameters, like the number of stages, the pump saturation of each individual stage and the input power. Unfortunately, all of these parameters increase the relaxation rate of the amplifier, so entirely constant pulse energies are typically not achieved. However, an optimized design of the amplifier chain allows for an effective damping of the pulse energy variation (Figure 5). Especially at high sweep-frequencies, a quite constant value is possible. The remaining variation can easily be actively compensated, if necessary.

Range of pulse repetition rates widens

A USP laser with an average power of more than 200 W with pulse duration of 40 ps or less, depending on the oscillator, has been realized. The PRR can be changed more than 30,000 times per second between 5 MHz and 10 MHz with a pulse energy variation of less than 10 percent without an active compensation. An even wider sweep range of the PRR between a few 100 kHz and 40 MHz and average output powers of more than 400 W are currently under development in a governmental funded project[7]. There, a new type of energy controller is also developed, which operates passively and further smoothens the pulse energy.

With the present system, ultrafast, on-the-fly micromachining of scattering centers for illumination applications have been demonstrated. More than 1.6 million dots per second were written in a polymethyl methacrylate plate with a reproducibility of 1 μm[8]. Further applications, especially with tight trajectories, where the advantages of the new type of laser system are obvious, are under development.

References

1. GFH GmbH (April 2013). UKP-Laser mit hoher Bahngeschwindigkeit, Mikroproduktion.
2. GFH GmbH (January 2016). Der Laserstrahl als Drehwerkzeug, Mikroproduktion.
3. BMBF, Erforschung und Entwicklung von innovativen hybrid-integrierten Diodenlaser-Komponenten und Systemen (INDILAS), 13N9817.
4. A. Klehr, et al. (2011). Compact ps-pulse laser source with free adjustable repetition rate and nJ pulse energy on microbench. *Proc SPIE*, Vol. 7953, No. 79531D.

5. F. Harth, et al. (2016). Ultra high-speed micromachining of transparent materials using high PRF ultrafast lasers and new resonant scanning systems. *Proc SPIE*, Vol. 9736, No. 97360N.

6. F. Harth, et al. (2016). Ultrafast laser with an average power of 120 W at 515 nm and a highly dynamic repetition rate in the MHz range for novel applications in micromachining. *Proc SPIE*, Vol. 9726, No. 972612.

7. Hochleistungs-UKP-Laser zur Mikromaterialbearbeitung mit variabler Pulsfolgefrequenz (HiPoRep), BMWi, VP2837408AB4.

8. F. Harth, et al. (2016). Ultra high-speed micromachining of transparent materials using high PRF ultrafast lasers and new resonant scanning systems.

Meet the authors

Florian Harth received his degree in experimental physics at the Technical University of Kaiserslautern in Germany. Since 2009, he has worked in the field of ultrafast laser source development and micromachining at the Photonik-Zentrum Kaiserslautern eV (PZKL).

Thomas Herrmann studied physics at the University of Kaiserslautern and received his Ph.D. degree in 1999. He was co-founder of the ultrafast laser company Lumera Laser GmbH (now Coherent Kaiserslautern) and was head of Lumera's application lab from 2003 to 2009. In 2009, he joined the PZKL and is responsible for the micromachining application center.

Bernhard Henrich studied physics and received his Ph.D. at the University of Kaiserslautern in 2000. He is co-founder of the Lumera Laser GmbH, and developed the first RAPID laser. In 2013, he joined the PZKL as technology manager.

Johannes A. L'huillier studied physics at the University of Kaiserslautern where he received his Ph.D. in 2003. His research is focused on optical parametric processes, ultrashort laser pulse, as well as on laser micromachining. Since 2009 he has served as the CEO of the PZKL.

Manufacturers Unite to Create Submicron Precision

When it comes to attention-to-detail, few industries compare with the medical sector, where flawless products are required. Here, the most precise devices can be life-changing — or even life-saving — for patients.

BY MARIE FREEBODY, CONTRIBUTING EDITOR

Ultraprecision manufacturing and optics unite two sophisticated technologies that are essential for a modern world that demands exact and meticulous results. From spacecraft components, next-generation displays and electronic devices to low-cost photovoltaic cells, defense and security technologies, ultra-precision machining is the catalyst behind many of today's frontline products.

The pioneers driving this area of industrial fastidiousness come from many different sectors. Photonics approaches are largely dominated by laser machining, but there are many other methods that can sometimes be an alternative or, as in an emerging trend, used alongside each other for superior results.

Excluding laser micromachining of the circuitry associated with medical devices, the main applications include creating stents for coronary, peripheral and neurological uses; cutting holes and other features in disposables, such as catheters and infusion/fluid delivery devices; and creating permanent counterfeit-proof

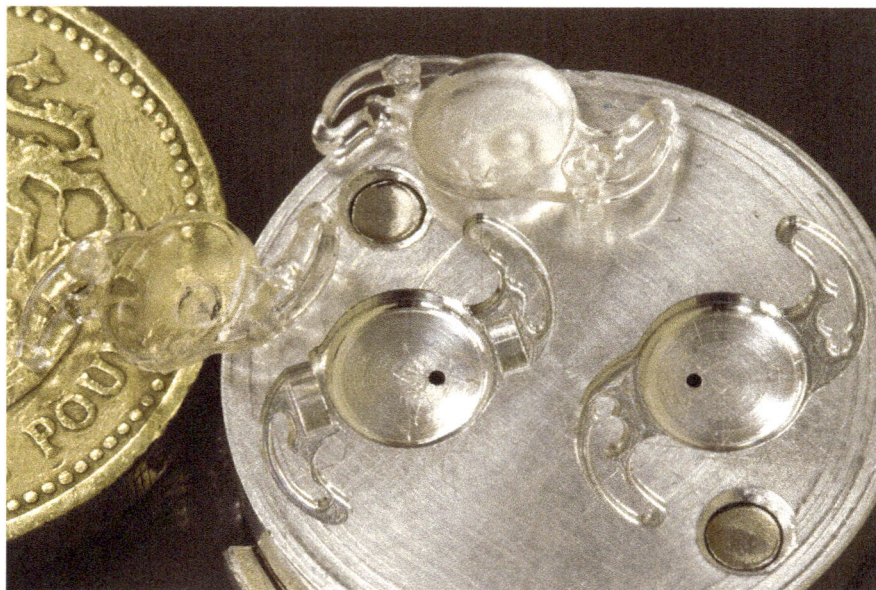

An ocular implant with a micro-fluidic dosing system that eliminates the need for post-operative medicine and/or targets the drug directly to the implant. Courtesy of the Centre for Innovative Manufacturing in Ultra Precision and Cranfield University.

marks without surface damage in plastic products, glass syringes and vials.

"Laser micromachining offers the potential for removal or transformation of material with incredible spatial selectivity and three-dimensional flexibility. It is therefore a superior tool for creating small features, and features and edges with smooth surfaces," said Joris van Nunen, product line manager for Industrial Ultrafast Lasers at Coherent Inc. "It's also useful for processing very thin materials, and materials that are delicate or thermally sensitive."

Cool cutting

Whether cutting with a laser or a diamond mill, the heat that is generated as a byproduct should be minimized. The goal is to achieve cold or athermal ablation — removing material with an absolute minimum of peripheral heating, which could manifest as burrs, slag or recast material. Incidentally, this is a common goal in industries such as microelectronics and solar, meaning the manufacturing progression is similarly mirrored.

"This heat-affected zone can be minimized by using shorter wavelengths

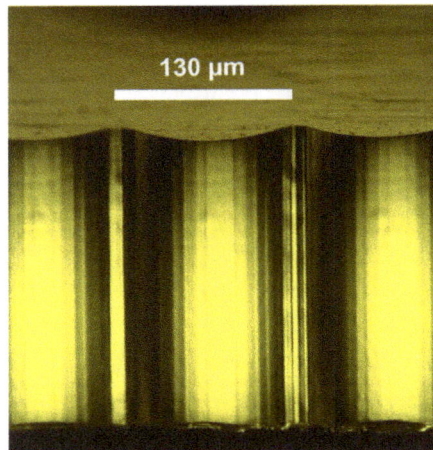

130 µm

A microscopic picture of a cylindrical micromirror array from diamond shaping. These mirrors are the functional areas of a mold for polymethyl methacrylate parts used in opto-electronic devices with up to 16 parallel waveguides for decoupling the light beams into free space toward the detection chamber with marked cancer cells. Courtesy of Kugler GmbH.

Detail of microfluidic mold insert. Left half shows a mixer and reaction area for two fluids, while the right half depicts an extended flat chamber for interaction of fluid constituents with light. Courtesy of Kugler GmbH.

and/or shorter pulsewidths," said Frank Gaebler, marketing director for materials processing at Coherent Inc. "For this reason, a few years ago, we saw fast growth in the use of excimers and visible nanosecond DPSS [diode-pumped solid-state] lasers; subsequently, there was a move toward ultraviolet DPSS lasers."

The next big development came in the form of industrial ultrafast lasers, which produce an even smaller heat-affected zone (HAZ) than nanosecond infrared or visible lasers. Here, femtosecond lasers almost universally deliver superior surface and edge quality compared with any previous industrial laser type.

"At first most of these operated in the picosecond regime, but today laser manufacturers now also offer femtosecond performance in industrial-grade platforms," Gaebler said. "The use of ultrafast laser pulses enables cutting of novel hole shapes and modalities in catheters, for example, and allows marking inside glass products, such as syringe bodies, with no impact on the surface, material integrity or the contents."

For any production line, process optimization is a challenge. In some cases this can be achieved quickly — a single day — but for some medical precision manufacturing applications, optimization can take a month or more. As laser pulse widths shorten and surface finish becomes just as critical as throughput and process economics, optimizing the process can mean the difference between success or failure.

When it comes to "routine" machining applications using a carbon dioxide laser, for example, process optimization can typically be completed in around a day in a laboratory or at the customer site. Even when using state-of-the-art nanosecond lasers, process optimization rarely takes longer than two days.

"But picosecond laser processes often require a whole week of iterative test runs and improvement," Gaebler said. "And with femtosecond lasers, some medical precision manufacturing applications can entail a month or more of careful optimization."

Ultrafast lasers pose another problem: In the past, the reliability of industrial femtosecond, and even picosecond, lasers was not optimal, often compromising 24/7 applications. What's more, these lasers were only available for a premium price, compared with nanosecond lasers.

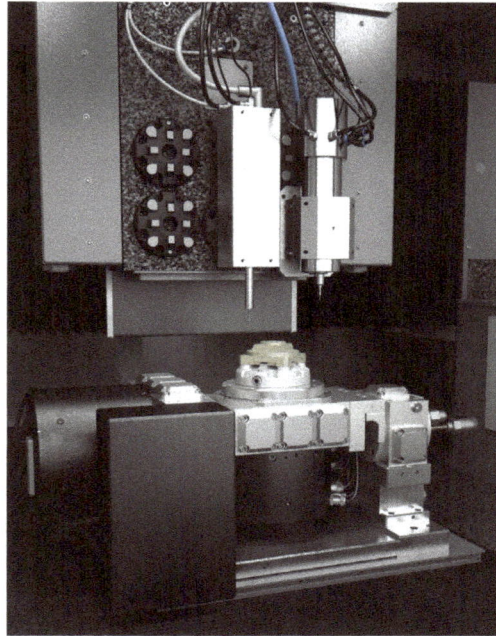

The Kugler MICROGANTRY nano3/5X is a high precision, CNC-controlled 3- to 5-axis machining center with aerostatic bearings, which is specially designed and optimized for the demands of micromachining and microstructuring with the use of microcutting and/or laser technology. The image shows a view into the machining space. The lower Y-axis slide carries a turn and swivel unit with the workpiece. It is spanned by the XZ-axes gantry where a spindle for micromilling and a co-linear touch probe measuring system are attached to the Z-slide. Courtesy of Kugler GmbH.

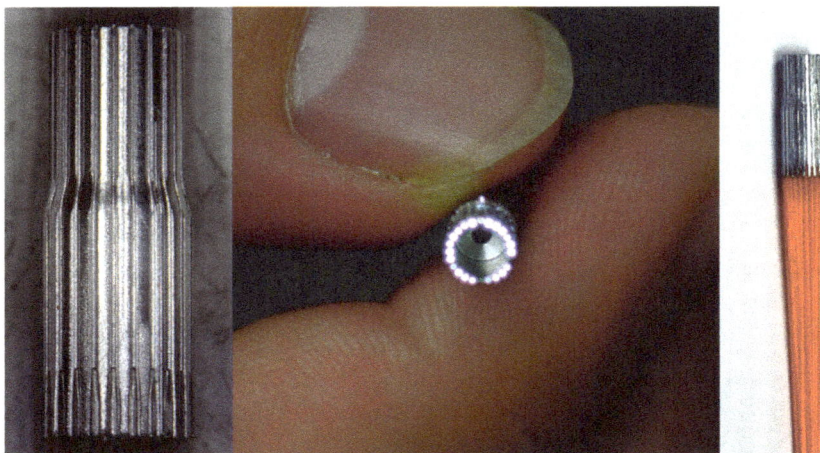

Micromachined fiber guide sleeve for earscanner. Optical fibers become embedded into the axial slots of the guide sleeve (left). Completed fiber guide with lighting fiber ends (right). Courtesy of Kugler GmbH.

Demonstrating fluid system mold for different lab on-a-chip functions. This is a feasibility study with several structures and geometry elements typical for biochemical fluid systems. Courtesy of Kugler GmbH.

While early adopters were willing to tolerate these issues, laser manufacturers knew they had to address the issues of process costs and laser reliability before the wider market would welcome the devices.

"In the last five years, we have seen tremendous advances in laser design, manufacturing and testing, and particularly the rigorous use of HALT/HASS testing practices and tools, long used in other nonphotonic industries," van Nunen said. "The result is a new generation of ultrafast lasers, such as the Coherent RAPID NX and Monaco family, which provide the same levels of reliability and lifetime as users routinely expect from other electronic instrumentation."

Highly accelerated life test (HALT) and highly accelerated stress screening (HASS) are methods of fast-tracking product reliability that are commonly used by the electronics and computer industries as well as by the military.

The hope is that as automated laser assembly and economies of scale help to push down the cost of femtosecond lasers, combined with levels of production throughput, high uptime and reduced cost of ownership, widespread adoption will be seen in many areas of medical manufacturing.

Molds, microfluidics and medical disposables

Many medical devices such as optical molds, lenses and fibers are machined using nonlaser-based methods such as milling and diamond turning. Medical technologies offer a substantial market for machine tool engineering, which is largely addressed by a number of well-known machine tool manufacturers in the south of Germany, including Kugler GmbH based in Salem, Germany.

"The most important application of precision manufacturing by high precision milling and micromilling in medical technology consists in the manufacture of mold inserts," said Klaus Baier, R&D project manager at Kugler.

Optical molds produce some of the smallest lenses and microlens arrays for optical instruments in medical analytics and diagnostics. Molds are also needed for large volume production of items such as lab-on-a-chip and microfluidic systems; multiwell and microplates; cell and tissue culture flasks; and products for blood, urine and specimen collection, which are increasingly used as medical disposables.

"Micromilling and microdrilling can be used for the direct preparation of PMMA [polymethyl methacrylate] fibers with diameters less than 500 microns and fiber components which are applied in sensor systems for the minimal invasive blood glucose measurement and in-ear scanners," Baier said.

The next frontier in medicine will take advantage of microsystems, which play a key role for in-vitro diagnostics and personalized medicine and health care. Implantable drug delivery systems and artificial organs that are currently under development rely on MEMS (microelectromechanical systems) and MOEMS (micro-optoelectromechanical systems) that are products of ultra-precise manufacturing.

Mold insert for microfluidic components. Courtesy of Kugler GmbH.

150.00 um/div

Pyramid microstructure for multiwell plate functionalization. This structure was created by diamond machining on the tips of a finger array in a mold for polycarbonate multiwell plates used in medical analytics. In combination with a superimposed nano-structure due to a special CVD-coating of the microstructured mold fingers, the bottoms of molded wells show the lotus effect for aqueous samples. Courtesy of Kugler GmbH.

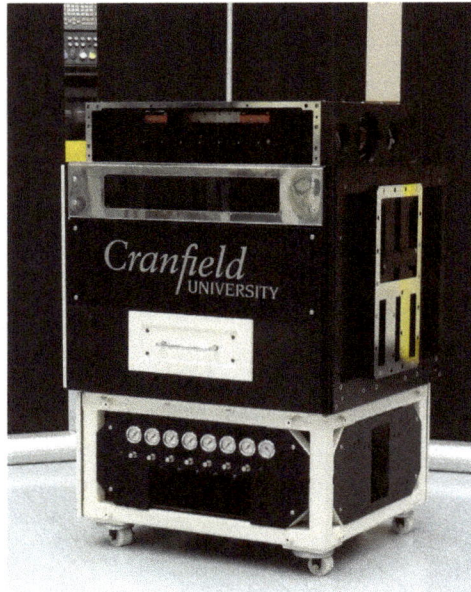

Micro-4 Machining from Loxham Precision Ltd., a 'micro-factory' precision machining center (Centre for Innovative Manufacturing in Ultra Precision). Courtesy of the Centre for Innovative Manufacturing in Ultra Precision and Cranfield University.

"Machining techniques including diamond turning/shaping and micromilling are a precondition for the manufacture of all kind[s] of microsystems which are indeed on the cusp to revolutionize medical engineering," Baier said.

Best of both

Competition is fierce in the manufacturing industry and champions of one approach may have once viewed each other as rivals. But things are changing, and an emerging trend is increasingly transforming competitors into collaborators.

"An interesting new market trend we are seeing at Coherent is increasing interest in lasers from companies specializing in various classical tools, such as EDM (electron discharge machining)," Gaebler said.

"EDM has long been used for drilling small holes and etching features, but is limited to metals or substrates precoated with metals. So lasers, and particularly femtosecond lasers, allow these tool builders to equip their systems with a material-neutral tool that provides 3D machining capabilities as a complement or alternative to their EDM technology."

Researchers at the U.K.'s Centre for Innovative Manufacturing in Ultra Precision based at Cranfield University in Bedfordshire predict the future of innovative manufacturing lies in hybrid machining centers that combine laser with ion beam figuring and plasma with beam figuring.

CAD drawing of the 1.4 m-wide Roll-to-Roll research platform under construction at Cranfield University (Centre for Innovative Manufacturing in Ultra Precision). Courtesy of the Centre for Innovative Manufacturing in Ultra Precision and Cranfield University.

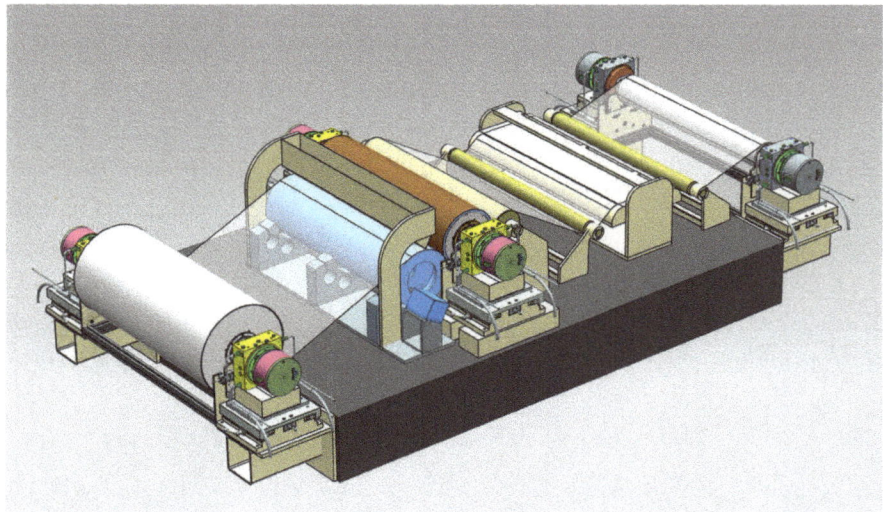

The result is an enhanced surface finish down to subnanometer levels or to 10 nm featuring faster than possible in conventional single processing centers.

"The development of miniature ultra-precision machining centers such as the Loxham Precision [based in Cranfield, U.K.] Micro-4 permit high levels of machining precision at higher volumes than could previously be achieved," said Martin O'Hara, national strategy manager for Ultra Precision at the Engineering and Physical Sciences Research Council-funded center.

"By using automated loading/unloading, maintaining the machining environment away from the user access and reducing overall moving mass to enable single-phase operation, parts can now be produced with submicron precision in almost any location," he added.

This enables customized micro-factory manufacturing to be located in retail environments for optical glass manufacture, for example, and clinical environments for ocular implants fabricated to patient requirements.

In another emerging trend, 3D printing offers some enticing benefits for creating reproducible and made-to-measure products for medical applications. But for all of its advantages, 3D printing cannot compete with laser and other traditional processes in terms of speed and cost of operation. "At the moment the methods of 3D printing do not yet yield the dimensional accuracy needed for all the parts described above," Kugler's Baier said. "Furthermore not all the biocompatible materials are available for 3D printing."

Again, the future seems to be to integrate the techniques. Experts at Coherent believe that combining novel laser-based additive manufacturing methods with the convenience of 3D printing concepts holds a lot of promise and potential market growth.

An emerging real world example of this involves metal sintering of titanium, which allows for bone (replacement) implants, for example, to replace a jawbone damaged in an automobile accident or removed because of cancer.

"3D scanning followed by 3D printing with a fiber laser enables creation of a perfect replica replacement that is unique and patient-specific," van Nunen said. "This is a great example of how precision laser machining can improve, rather than simply replace, some existing manufacturing methods."

For Glass and Silicon Wafer Cutting, Shorter Pulse Widths Yield Better Results

Micromachining produces clean cuts and marks with high surface quality and minimal heat-affected zone.

BY DIRK MÜLLER, COHERENT INC.

In manufacturing of semiconductor microelectronics, displays, medical devices and many other industries, there is an increasing trend toward higher-precision processing. This means cutting, drilling and marking parts with smaller feature sizes and greater accuracy, superior edge quality and with a reduced effect on surrounding material. In the past, most precision laser-based processing applications relied on nanosecond pulse widths or ultraviolet output (or both). But traditional sources cannot always service this new, demanding class of applications. As a result, some applications are now turning to lasers with ultrafast — picosecond or femtosecond — regime pulse widths to accomplish these tasks.

Ultrafast processing benefits

The goal of micromachining is the creation of micron-scale features, such as holes, grooves and marks, with high-dimensional accuracy while avoiding peripheral thermal damage to surrounding material. In other words, precise, clean cuts and marks with high surface quality and minimal heat-affected zone (HAZ).

There are two basic mechanisms by which a laser can precision drill, scribe, cut or mark a material. Many traditional applications rely on infrared and visible Q-switched lasers, which have pulse widths in the tens of nanoseconds range, and which remove material via a photothermal interaction. Here, the focused laser beam acts as a spatially confined, intense heat source. Targeted material is heated rapidly, eventually causing it to be vaporized — essentially boiled away.

The advantage of this approach is that it enables rapid removal of relatively large amounts of target material. Furthermore, nanosecond laser technology is mature; these sources are highly reliable and have attractive cost of ownership characteristics. However, for the most demanding tasks, peripheral HAZ damage and/or the presence of some recast material can present a

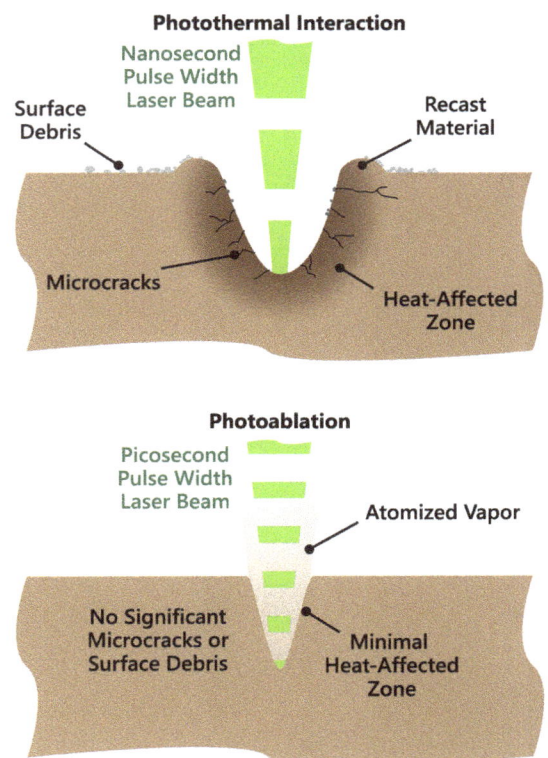

Schematic illustrating the major differences between ultrafast processing and processing with longer pulse width lasers. Courtesy of Coherent.

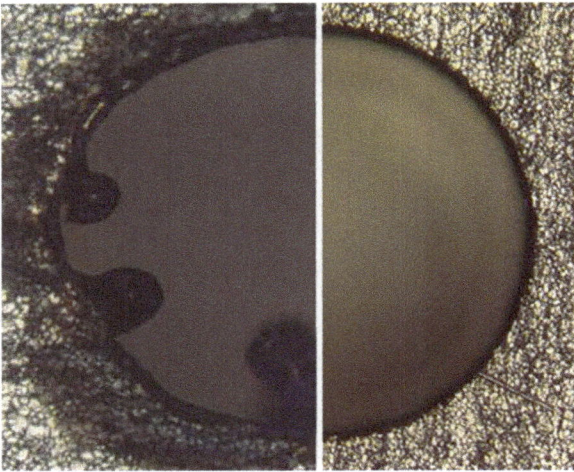

Comparison of a 200-µm-diameter hole drilled in stainless steel with a nanosecond laser (left) and a picosecond laser (right). The picosecond laser produced a cleaner hole with less recast material and a smaller heat-affected zone (HAZ). Courtesy of Coherent.

limitation. This includes the delamination of surface coatings, microcracking, or changes in the bulk material properties.

One way of minimizing the size of the HAZ is to employ a nanosecond laser having output in the UV, rather than in the visible or near infrared. UV light is strongly absorbed by most materials, limiting how far the laser light penetrates into the part and therefore reducing the HAZ.

The second mechanism for laser material removal is based on photoablation, which involves directly breaking the molecular or atomic bonds that hold the material together rather than simply heating it. This can be performed with ultrafast lasers because their short pulsewidths lead to very high peak powers (megawatts and above). These high peak fluences drive multiphoton absorption, which strips electrons from the material and then explodes away because of Coulomb repulsion. When using ultrafast pulses, the material is exposed to the laser energy for such a short time that the energy can't be carried beyond the area of impact, so the surrounding area stays cold. Energy left over after the bond-breaking process is carried away with the expelled particles. Together, these effects result in an inherently colder process with significantly reduced HAZ. This is also a very clean process, leaving no recast material and thereby eliminating the need for elaborate post-processing.

Another major advantage of ultrafast processing is that it is compatible with a very broad range of materials, including several high-bandgap materials including glass, sapphire and certain fluorinated polymers that have low linear optical absorption and are therefore difficult to machine with existing commercially available lasers. More specifically, the technique is "wavelength neutral"; that is, nonlinear absorption induces a light-matter interaction even if the material is normally transparent at the laser wavelength.

One limitation of ultrafast processing is that it provides lower material removal rates, and ultrafast lasers have been more costly than long pulse laser sources. As a result, ultrafast processing is typically reserved for applications that demand the greatest possible precision, quality and smallest HAZ.

From sapphire cutting to LED dicing

Because of their advantages, ultrafast processing lasers have now been adopted in a number of demanding, high-precision applications. These include LED dicing; sapphire cutting; drilling of automobile engine fuel injection nozzles and engine cooling plates; hole drilling and structuring of biomedical filters; cutting and drilling of FR-4 resin; cutting and drilling of both low-temperature co-fired ceramics and high-temperature co-fired ceramics; and microprocessing of metals such as stainless steel and copper. More recently, picosecond lasers have also been used to mark integrated circuit packages. The short

Results of stealth dicing a silicon carbide wafer with a picosecond laser, before mechanical separation of the individual dies. Courtesy of Suzhou Delphi Laser Co.

pulses ameliorate potential damage to the embedded die during marking.

Several Asian tool manufacturers have been using Coherent RAPID series picosecond lasers for several years now, specifically for semiconductor wafer dicing and glass cutting. Joshua Zhao, sales manager at Suzhou Delphi Laser Co. for the Americas region, discussed how the lasers are employed and the benefits they have delivered. "Wafer dicing can actually be accomplished in two different ways. In the first, called laser grooving, the beam is focused onto the surface of the wafer in the street area — the empty area between circuit components," he said. "The laser makes a scribe part way through the wafer, and the individual chips are subsequently singulated mechanically. The second method is called stealth dicing. Here the beam is focused within the wafer itself, where it creates a scribe within the material. Again, the individual chips are then singulated mechanically, typically by tape expansion.

"Previously, we employed a 355-nm nanosecond laser for laser grooving, but now we've switched to a 1064-nm picosecond laser and perform stealth dicing. This has delivered several benefits. First, the picosecond laser produces a much smaller heat-affected zone than the nanosecond laser," Zhao explained. "This allows the cutting street to be reduced in size from 25 µm, down to 14 µm. Which, in turn, results in a higher yield — that is, one can pack more devices onto a given wafer. Also, we have fewer pieces that don't separate properly than before, generating less waste. As a result, we're also able to run the process faster than before. For example, the nanosecond laser could process 15 wafers per hour, for a 10 mil × 23 mil chip size, but the picosecond laser can process 23 wafers per hour. Plus, we can even process thicker wafers; now we can dice 200-µm thick wafers, while our nanosecond laser process couldn't go above 100 µm thickness."

Cross sections of glass cut with a nanosecond laser (top) and a picosecond laser (bottom). The picosecond laser produced a cut with fewer microcracks and less residual debris. Courtesy of Suzhou Delphi Laser Co.

Picosecond laser glass cutting: avoiding cracks and debris

Another important picosecond processing application is glass cutting. This application is driven by the tremendous market growth for cellphones and tablet computers that incorporate touchscreens. There are two important trends in touchscreen display glass scribing. The first is a drive toward the use of thinner glass substrates in order to minimize the total weight of the display. The second is a need to cut curved shapes in the glass, rather than simply straight lines, in order to allow rounded edges on the display, as well as to accommodate more complex screen geometries.

As the display glass gets thinner, it is critical that the finished product still retain the ability to withstand being dropped, handled roughly and pressed upon (for

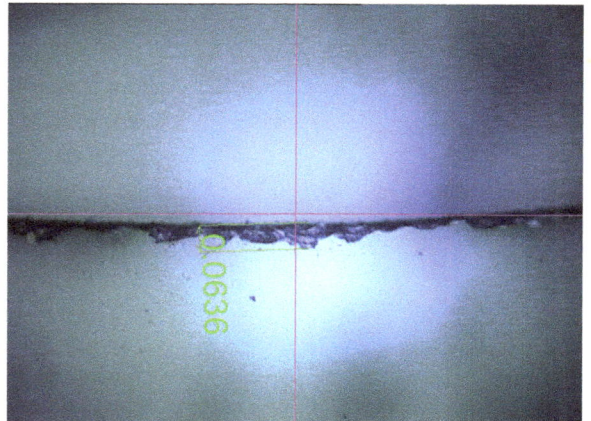

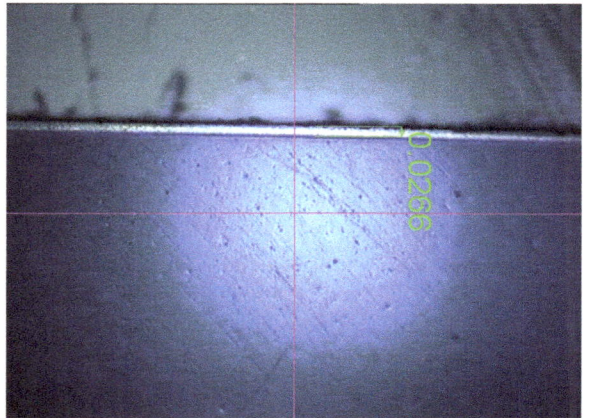

touchscreens). A typical LCD touchscreen actually contains three or four stacked layers of glass. The topmost (outer) sheet is often a 700-µm-thick protective cover glass. To minimize the risk of scratching and breakage, this outer layer of the topmost glass is chemically treated to produce a very tough surface — Corning's Gorilla Glass, Asahi's Dragontrail and Schott's Xensation are examples of this. The thickness of this strengthened layer typically runs tens of microns deep.

Traditional mechanical glass cutting can lead to microcracks and debris. Laser glass cutting, based on CO_2 and nanosecond solid-state lasers, has been in use for some time in the display industry. Both of these lasers produce dramatically better results than mechanical cutting, but can each have some limitations, especially for very thin glass (<300 µm thickness). Zhao said, "[Picosecond laser cutting] yields a finished piece that has greater edge strength, and is much more resistant to breaking during use."

A new generation of reliable, high-power, industrial ultrafast lasers is enabling higher-precision microprocessing in a variety of applications. Expect this technology to impact industries as diverse as microelectronics fabrication, medical device manufacturing and automotive production.

Meet the author

Dirk Müller, Ph.D., is the director of Strategic Marketing at Coherent Inc.

Cut, Mark, Drill, Repeat

Laser machining drives improvements in medical devices.

BY LYNN SAVAGE, FEATURES EDITOR

Increasingly, the medical devices used to support good health and provide emergency and long-term care are not being designed to sit on a rack or table next to a patient's hospital bed. Instead, they are being made small and light enough to be clipped onto clothing, to be worn like a bracelet or badge, or to be implanted directly into the body. Ultimately, the aim of medical device design is to provide doctors and nurses with better tools that are always within reach, to improve the overall health of patients with treatments that have fewer side effects, and to keep people out of hospitals altogether, thus reducing health care costs.

Some of the most common medical devices today are sensors that detect and read out biometrics such as blood pressure or sugar/insulin response and implants such as stents for helping keep open clogged or damaged arteries or tissues.

Once upon a time, medical device manufacturing consisted mostly of stamping

Systems such as Rofin-Baasel's Starcut Tube (inset) are used to manufacture precision-cut parts, including slender cylinders for stents and catheters. Courtesy of Rofin-Baasel.

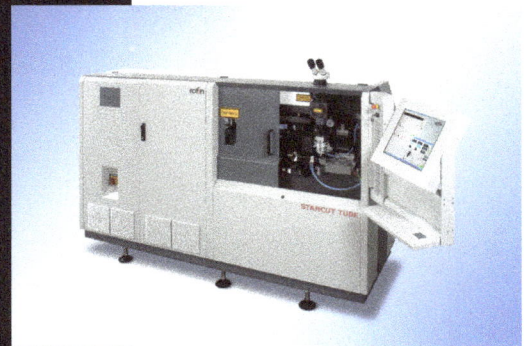

Laser-based manufacturing can perform nearly all of the tasks required to build medical devices, including drilling, structuring, cutting, trimming and marking. Courtesy of Laser Micromachining Ltd.

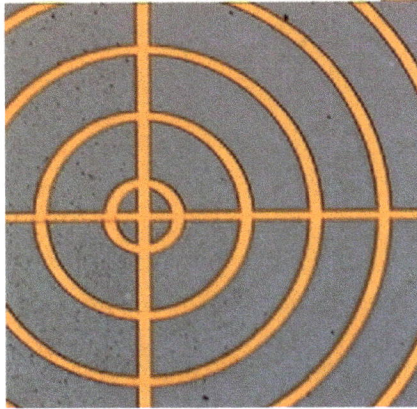

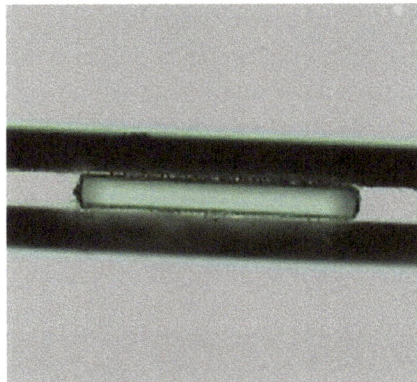

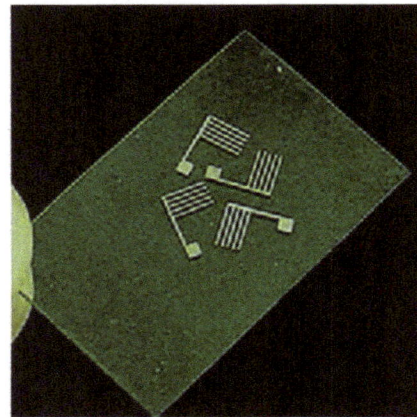

sheet metal and pouring molten metal into molds. This type of manufacturing has not gone away, but it is being supplanted by new techniques — such as laser-based machining — that promise to achieve two chief goals: reducing the size of the apparatuses and increasing the use of novel materials inside them.

Efforts are underway to design sensors that can closely track health-related patient biochemistry and the effectiveness of pharmaceuticals and other treatments. Similarly, stents, novel artificial bone and other tissues, and inserts such as catheters and artificial retinas are continuously being redesigned. To make these sensors and inserts smaller, stronger and more effective, materials such as the shape-memory titanium-nickel are being tested and used. Manufacturers are finding that lasers can effectively work a wide variety of novel materials.

Contract manufacturers such as Resonetics LLC of Nashua, N.H., work with a multitude of materials, including polymers, glass, ceramics and ultrathin sheets of metals and alloys. They drill, cut, scribe and mark tiny device components and turn them into catheters, stents, dental implants and biosensors that help track diabetes and heart health. A relatively simple catheter once took weeks of tooling by hand or with large industrial machines, but now it can be created in a fraction of the time, using lasers to prepare the major components and finish the final product.

According to Nadeem Rizvi, co-founder and managing director of Laser Micromachining Ltd. in St. Asaph, U.K., polymers and thin metals are the core materials used in most medical devices, but they don't paint the whole picture.

"There seems to be a trend toward thinner materials with finer features and also a growing demand for devices with precision structures in combinations of materials (e.g., polymers on glass or thin coatings on different base materials)," he said. "There is also quite a lot of interest in completely new materials, such as biodegradable polymers for implants and tissue-engineering applications."

Drill, laser, drill

Among the multitude of steps that comprise medical device manufacturing, the key ones are forming the basic shape of each part; drilling or punching holes, notches and other shapes where needed; attaching the parts together; and marking all or some of the parts for later identification. Where large mechanical presses once were used to stamp out shapes from ¼-in. sheet metal like a cookie cutter, lasers now can cut and trim ultrathin sheets of metals and polymers to very close tolerance. The difference between techniques has enabled metal tubes to shrink to such a small size that stents are now ubiquitously used in patients' arteries, ureters or similar tubular structures.

Similarly, laser-based welding, ablating and structuring, and marking or scribing all take the place of traditional manufacturing methods, while improving the form and function of the final product.

Drilling might be the most common manufacturing step improved upon by laser machining. Tiny holes in many medical appliances — for example, the via holes that transport fluids from one chamber to another in a portable sample analyzer — can be made microns in diameter and millimeters long. Traditional drilling techniques cannot match the performance of lasers in this regard.

Sintering, a process in which a material in powdered form is shaped and turned into a solid part via applied external heat, also is being matched up with lasers. Laser sintering, the 3D equivalent of 2D laser printing, can create medical components of very tiny size to very small tolerances.

"Lasers are quite a flexible tool, especially compared with stamping and other manufacturing technologies," said Dieter Mairhörmann, international sales manager (medico) for Rofin-Baasel of Starnberg, Germany. "Fiber lasers and CO_2 lasers are common for sintering."

Prototyping

Besides fulfilling the steps required for manufacturing mass quantities, the other major service offered by laser micromachining companies is rapid prototyping — the creation of model components (or whole devices) in time frames typically shy of a week. This service offers the ability to audition several design options for a single device idea without large expenditures of either time or cash. Designing and building prototype medical devices previously took much longer, discouraging the pursuit of either big new ideas or subtle design tweaks.

"One of the main benefits of using laser machining for product development," Laser Micromachining's Rizvi said, "is that different design ideas can be tried quickly and the results used to develop the most effective product solution."

Norman Noble Inc. of Highland Heights, Ohio, is one of several companies that offer prototyping services, working with materials ranging from nitinol to thin metals to bioabsorbable polymers. Companies such as Norman Noble and Laser Micromachining provide the manufacturing prowess, while laser makers provide the base technology.

Mairhörmann said that implant makers and other medical technology clients

> 'The standard requirements of high quality, high tolerances and repeatable results have not changed and are not likely to.'
>
> *– Nadeem Rizvi,*
> *Laser Micromachining Ltd.*

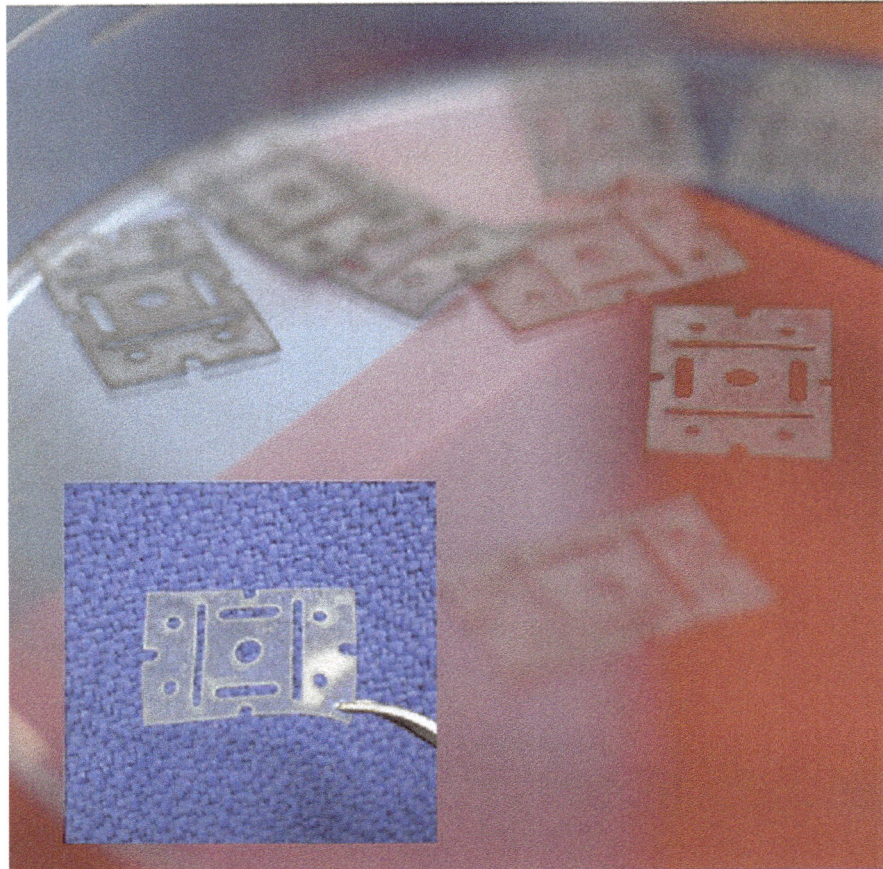

A wide variety of lasers are available to manufacturers, meaning that materials ranging from ¼-inch sheet metal to thin-film polymers can be processed to precise tolerances. Courtesy of Laser Micromachining Ltd.

are fairly knowledgeable about lasers because they know the parameters required. The manufacturers' customers — the device designers — are, on the whole, still catching up with the possibilities of laser machining for both prototyping and mass production.

"Clients who are not so aware of laser technology nonetheless are very quick to appreciate the benefits of laser machining, once they see how it can help their businesses quickly and cost-effectively," Rizvi said. "Still, from time to time, we will get an inquiry about drilling a 1-µm-diameter hole through 10 mm of metal, and we have to explain that lasers can't do everything."

Next up

Despite the learning curve that some device designers still must experience, laser machining equipment and services are largely commoditized. No sea changes are expected. Rapid prototyping services likely will increase, especially as new powderized base materials are developed. Rizvi and others believe that sintering and structuring — using lasers to shape or deposit materials into 3D parts — also will gather steam. Increasingly, innovations will be coming from startups and other small, nimble companies.

According to Terry Young, Julie Eatock and Dorian Dixon in a recent article in the *Journal of Manufacturing Technology Management* (Vol. 20, Issue 2, pp. 218-234), the majority of innovative, new-to-the-world medical devices are being developed by small companies. To the authors, large, more established companies seem complacent, spending their energies on incremental upgrades to existing devices. And these derivative products are less likely to exceed expectations of success than are novel items.

Technological improvements likely will be incremental, especially concerning lasers themselves. The use of fiber lasers likely will continue to increase, while wavelengths will shrink further into the ultraviolet as customer demand for finer, smoother features increases. Because shorter laser pulses mean that less heat is transferred to the material, femtosecond lasers will get a close look as successors to picosecond systems.

In all, the growing demand for better and more widely available health care in the U.S. and globally will drive the design and manufacture of small devices that preserve, protect and monitor the health of nearly everyone. Those who make lasers and those who use them to make medical devices have a long road of success ahead of them.

Picosecond Lasers for High-Quality Industrial Micromachining

Focused picosecond pulses are well-suited to avoid major thermal side effects and reach a new level of machining quality.

BY DIRK MÜLLER, LUMERA LASER GMBH

Just as CW and quasi-CW lasers have revolutionized the materials processing world, picosecond lasers are poised to change the world of micromachining. The use of picosecond lasers in micromachining was ushered in more than three decades ago because various millisecond and, later, nanosecond lasers had proved that pulsed lasers offer new capabilities.

Today, laser pulses in the millisecond to nanosecond regime are being used for drilling holes in wafers, cutting thin sheet metal and scribing ceramics, and for marking automotive parts, credit cards and passports. In micromachining applications, lasers outperform mechanical tools because of their flexibility, reliability, reproducibility, ease of programming and lack of mechanical force or contamination to the part.

Until the introduction of picosecond lasers, all applications were based on the principle of very fast local heating, melting and vaporization of the target mate-

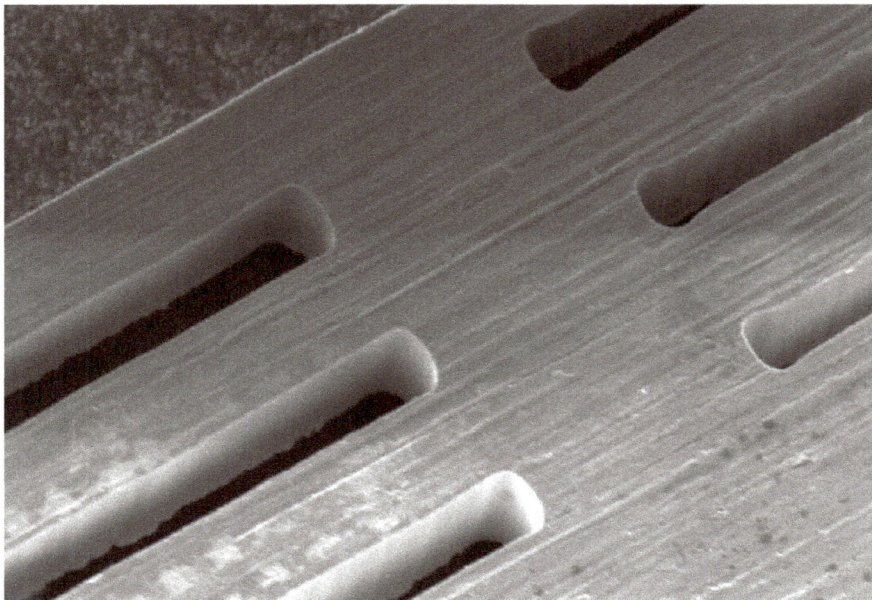

These 15-µm-wide slots are in a 20-µm silver sheet. All images courtesy of Lumera Laser GmbH.

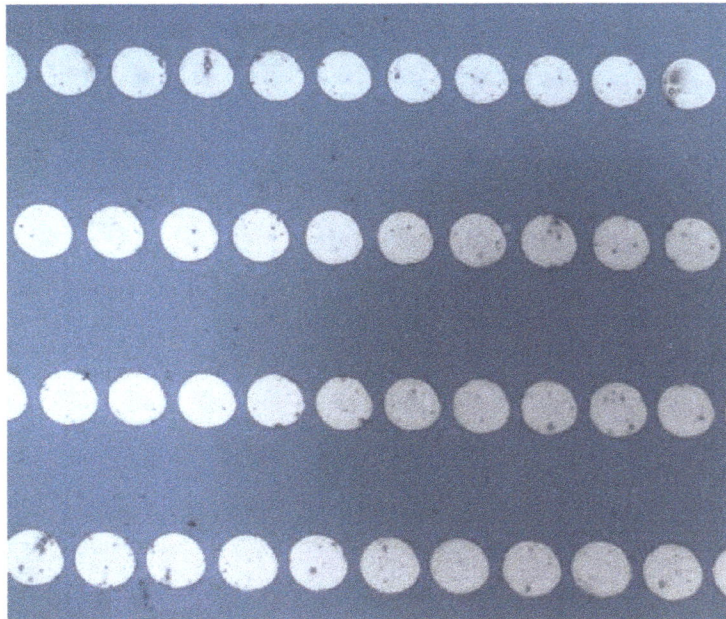

Shown is the removal of a 70-nm SiN layer on silicon, with up to 1 million dots per second. Dot diameter is 50 μm.

rial, which later caused thermal side effects such as burrs, recrystallization and microcracks in a product.

Industrial ultrafast lasers have become available recently. Focused picosecond pulses of appropriate energy are well-suited to avoid major thermal side effects and reach a new level of machining quality.

Perhaps equally significant is that picosecond pulses are material-unspecific, making them the universal machining tool because micromachining with picosecond pulses does not rely on an absorption process. Underlying the micromachining is the formation of a surface plasma cloud. The large electric field of a picosecond pulse strips the low mass electrons off the atoms, and the positively charged atoms left behind undergo a Coulomb explosion.

It could be argued that even shorter laser pulses in the femtosecond range would further enhance the micromachining quality. However, femtosecond lasers are significantly more complex, prone to failure and deliver less average power at a much higher price. The electric fields generated in picosecond laser pulses are sufficient to trigger this process, and going to shorter pulses only complicates the system. The high peak power of femtosecond pulses requires careful beam delivery and is prone to unwanted nonlinear effects.

Laser pulses in the picosecond range seem to hit the sweet spot: Picosecond pulses possess excellent beam quality, deliver the right pulse energy level and can be produced in a reliable, industrial package. They reach megahertz repetition rates with more than 50 W of average power, enabling economic industrial scale throughput. If the energy density is slightly above the ablation threshold ($\sim$1 J/cm^2), most materials will show an ablation of a 10- to 100-nm-thick material layer. The ablation threshold varies only slightly with the type of material (0.1 to 2 J/cm^2) and is mostly independent of wavelength, pulse length or other conditions.

Most micromachining applications are interested in sculpting a surface structure; i.e., the "cold" micro removal of material to scribe a trench, cut a shape, drill a hole, reveal a material layer or isolate an area.

Mechanical drilling/milling and electron discharge machining are overstretched in reliability and cost when creating structures <50 µm. A high-quality laser beam with an M^2 <1.5 can be conveniently focused to a 5- to 50-µm spot size, enabling machine structures of similar size. A typical focal spot diameter of 25 µm will require ~10 µJ of pulse energy to satisfy the 1-J/cm^2 ablation threshold criteria.

Higher pulse energy densities will not necessarily work better or faster: The ablated plasma cloud gets denser and can no longer dissociate from the surface. Even worse, it may thermally alter the material and destroy the "cold" quality. An ideal industrial picosecond laser source for "cold" micromachining should produce pulse energies in the range of 10 to 50 µJ at a repetition rate of ~1 MHz. At even higher repetition rates, a shading effect from the plasma cloud of the prior pulse eventually reduces the efficiency.

Recently, it has been observed that several picosecond pulses with nanosecond-scale separation (burst mode) not only will improve the ablation rate substantially, but also the micromachining quality; e.g., the surface roughness of blind holes.

Experiments with 50-W picosecond lasers operating in burst mode have achieved ablation rates of ±60 mm^3/min. Applications with low aspect ratio (depth/diameter) yield in glass up to 20 to 60 mm^3/min; in stainless steel, 10 mm^3/min; in silicon, 30 mm^3/min; and in organics and biomaterials, up to 10 mm^3/min.

The cost per photon has dropped by a factor of 10 within the past four years, making the picosecond laser a very economical choice and one capable of competing with many other machining choices on a cost-per-part basis. Although the initial investment for a picosecond laser of adequate power and beam quality is

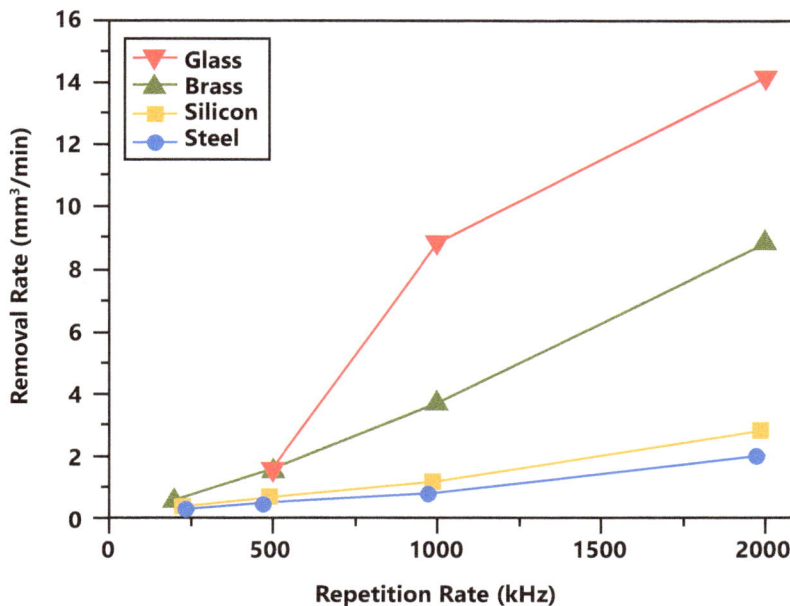

Removal rates increase with repetition rate.

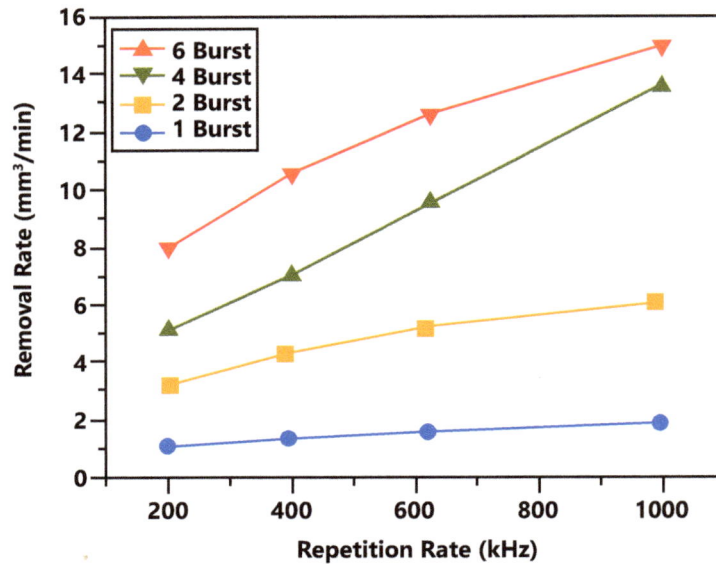

Removal rate substantially increases with burst mode operation.

higher than that for a nanosecond one, the total cost of ownership is only about $0.29/min. Within a minute, 20 mm^3 or more of any hard or difficult-to-machine material, such as cubic boron nitride or diamond, sapphire, glass or ceramics, can be removed.

An ever-growing number of interesting applications are emerging, where removal of thin, small volumes creates high value in the product: the semiconductor industry (low k materials); photovoltaic industry (especially thin-film technology); display technology (transparent conductive oxide, organic LEDs); micromolds on demand; and precise apertures and electrode structures, large microstructured surfaces for the printing industry, or embossing of rolls or medical implants.

Even large marine vessels are candidates for micromachining. Micron-size features offer antifouling protection and reduce friction. Further, injection nozzles for high-compression cylinders as well as cutting and drilling of thin glass materials are emerging as important high-volume applications. To cater to large-scale industrial applications, picosecond lasers are required to demonstrate more than one year of mean time to failure and <1 h of time to repair. The newest generation of industrial picosecond lasers already demonstrates such performance.

Analysts agree that the market for industrial picosecond lasers will grow tenfold within the next three to four years. Power levels of 500 W will be reached, costs will be reduced and throughput will be increased. Reliability and cost of ownership will be the most decisive metric for any large-scale industrial use.

Meet the author

Dirk Müller is director of business development at Lumera Laser GmbH in Kaiserslautern, Germany.

Surface
Treatment

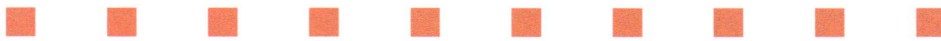

Testing the Limits of Excimer Lasers: Annealing for Advanced Displays

Mass production of high-resolution displays utilizes ever-larger substrate panels, placing unique demands on the excimer laser systems used in their production.

BY RAINER PAETZEL and RALPH DELMDAHL, COHERENT INC.

Low-temperature polycrystalline silicon (LTPS) is increasingly used as the thin-film transistor material on the glass backplanes of high-performance displays, particularly for smartphones. These thin films are fabricated on large glass panels that then are singulated into hundreds of individual screens. Mass production of LTPS on these panels is uniquely enabled by excimer lasers, moreover excimer lasers with extremely high pulse energies. These high pulse energies are needed in order to reach the requisite high process threshold intensity over a large area.

There is a fast-growing interest in extending LTPS to process larger area panels for several reasons: greater economy of scale, better and brighter mobile LCD screens, and the adoption of active-matrix organic light-emitting diode (AMOLED) smartphones and tablets. But as processes evolve to support ever-

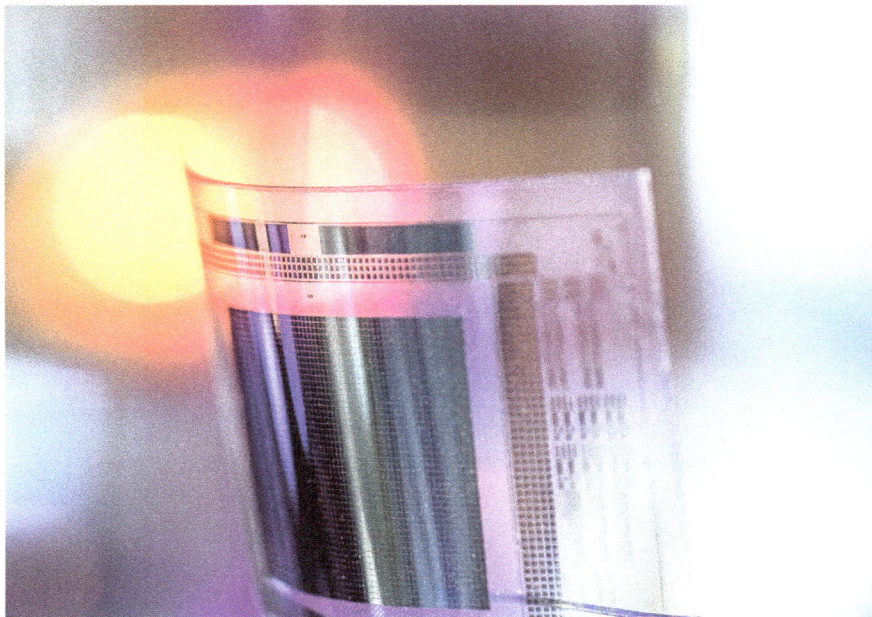

Excimer lasers enable high-volume manufacturing of flexible displays; a key step is laser lift-off where high-duty cycle excimer lasers are used to release the flexible displays from a rigid temporary carrier. All images courtesy of Coherent Inc.

High-power scaling for large format poly-Si annealing shown here with two dual-oscillator VYPER lasers, with all four beams merging into a 1300-mm length processing beam.

larger panels, new requirements are placed on the ultrapowerful excimer systems and the associated beam delivery and beam shaping optics used in the process. In what follows, the authors examine why excimer systems (laser, optics and internal diagnostics) are using modular architecture to deliver higher and higher energies with improved pulse-to-pulse stability and beam uniformity. For display manufacturers this means faster process throughput (screens per minute) and even better process consistency.

Polycrystalline silicon vs. amorphous silicon

The backplane of both active-matrix liquid crystal displays (AMLCDs) and AMOLED displays can be regarded as a type of large-scale integrated circuit, which starts as a 50-nm-thick film of amorphous silicon deposited on a thin glass sheet. For both display types, this is transformed into a network of thin-film transistors (TFTs) to enable each subpixel (red, green or blue) to be individually addressed.

Solid silicon can exist in three forms: single-crystal silicon, amorphous silicon (a-Si) and polycrystalline silicon (poly-Si). In single-crystal silicon, all the atoms are arranged in one giant, extended regular lattice. Large area, single-crystal silicon — as used in integrated circuit (IC) chips — is neither economically viable nor necessary for display applications. Rather, a-Si or poly-Si is used.

In amorphous silicon, the atoms are irregularly located with a high degree of disorder, resulting in the lowest electron mobility. In polycrystalline silicon, the atoms are arranged in microcrystals or grains, with discontinuities between the grains. The atoms can be highly ordered at the microscopic level, for instance, but disordered at the macroscopic level. This type has intermediate electronic properties, with the electron mobility and other parameters being highly dependent on the grain size and degree of order (e.g., grain size uniformity).

For AMLCDs, the choice is a trade-off between performance and cost: The higher electron mobility of poly-Si maximizes circuit efficiency, enabling small-

Adapting Laser Lift-Off Technique for Flexible Displays

Excimer lasers optimized for high-energy pulses have applications beyond annealing low-temperature silicon, including mask-based direct patterning of microcircuits for cost-sensitive applications like medical disposables, where every year 20 billion disposable sensor circuits are created in thin metal on a flex substrate in a reel-to-reel process. Another is laser lift-off for next-generation flexible displays.

Laser lift-off with ultraviolet lasers is well-established in microelectronics. As an example, consider blue laser diodes, where the circuitry has to be created on, then removed from, a durable carrier substrate

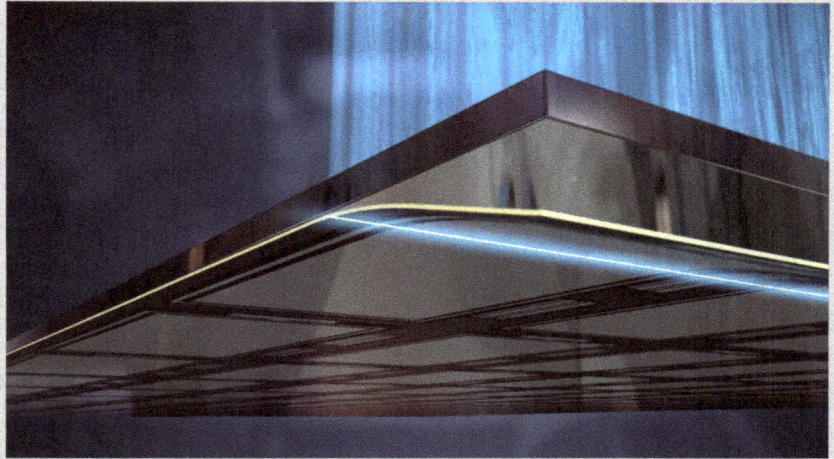

The transition from rigid to flexible display manufacturing is accomplished by separating the glass panel using an excimer laser line beam.

such as sapphire. The challenge with flexible displays is the sheer size of the product, which is hundreds of times larger than a typical IC chip and needs correspondingly higher laser power to complete the lift-off in an economically practical time.

In flexible display production, a glass panel that serves as temporary carrier for handling purposes is first coated with a polymer film. Display circuitry is then created on top of this layer. Finally, laser lift-off accomplishes the transition to full flexibility by passing an excimer laser beam (one pulse per area) through the glass carrier and vaporizing the top few atomic layers of the polymer. With the appropriate excimer laser line beam system, a Generation-4-size glass carrier (730 × 920 mm) containing 60 five-in. flexible displays is rapidly and gently separated applying only a few thousand laser pulses within seconds.

er area TFTs and narrower circuit traces. Reducing the size of these backplane components minimizes blocking of the backlight (which passes through this circuitry), making the display more efficient. In fact, poly-Si is key when it comes to an AMLCD with a pixel density of more than 300 ppi. This holds true for most of today's smartphones, where the smaller screens and short viewing distances necessitate higher pixel densities (between 400 and 900 pixels per inch) to provide a crisp and brilliant appearance of the screen. Plus, electrical efficiency is even more critical in portable devices as it directly influences battery life.

AMOLED pixels emit light directly, so there is no backlight to block with control circuitry. However, stable, low impedance control circuitry is critical because each pixel draws current, making electron mobility and TFT current stability even more important for these displays. This makes poly-Si the only choice, and it is hence used for all small- to medium-sized AMOLED displays.

As processes evolve to support ever-larger panels, new requirements are placed on the ultrapowerful excimer systems and the associated beam delivery and beam shaping optics used in the process.

Reshaping the output beam for a narrow profile

When silicon is vapor deposited on the glass backplane, it naturally assumes the amorphous form. High-temperature annealing to create poly-Si is not viable as it would require temperature-stable panels made of expensive quartz. Therefore, the key enabler for low-cost, mass production of displays using conventional display glass was the 308-nm excimer laser for selective low-temperature polysilicon recrystallization. The excimer laser process to form poly-Si is commonly referred to as excimer laser annealing (ELA).

In ELA, the excimer output beam is reshaped into a long narrow line profile. The length of the line ideally matches either the half or full width of the panel; hundreds of smartphone displays are eventually diced from each panel. The panel is translated under the line beam inside an ambient controlled chamber, ideally in a single-pass or in a two-pass scan process. The panel motion is matched to the 600-Hz laser repetition rate so that every part of the panel is irradiated by about 20 pulses.

The microcrystalline structure of the polysilicon determines the electron mobility and, hence, is very important to the quality of the final display. In turn, it is controlled by the laser pulse parameters, including the pulse energy, the number of pulses per location, the line beam profile (along both the long and short axes), the temporal profile of the laser pulse, and the pulse-to-pulse energy stability. For these reasons, display manufacturers buy a line beam rather than a laser; they need a complete system that delivers the precise line beam parameters their process requires. This necessitates a high degree of beam delivery system expertise by the laser vendor, including multiple levels of internal diagnostics and system monitoring at every stage of the enabling beam path.

Multi-oscillator = high-power excimer

Although LTPS is well-established in mass production, it is a very dynamic field. From the perspective of a laser beam supplier, the key trend is to support ever-larger panels. Specifically, the challenge is to provide a line beam profile that matches the width of these larger panels without sacrificing intensity (fluence) or any of the other system performance parameters that affect the quality of the final poly-Si.

The lasers used in ELA are the most powerful excimer lasers ever commercially developed, with pulse energies as high as 2 joules. This was a good match for creating a line beam profile up to 750 mm in length as needed for Gen 5 panels. But, soon a higher power level was needed to support the step up to Gen 6 panels measuring 1500 × 1850 mm. Moreover, a strategy was needed to scale to even larger panels in the future.

It is simply not practical to make the individual laser oscillator larger and longer ad infinitum. Scaling issues like cooling, discharge uniformity, control of beam-quality, etc., preclude this. Instead, we chose a path of combining multiple lasers in a multi-oscillator concept, which resulted in VYPER, a very high-power excimer laser platform. And rather than maintaining the existing performance levels, this is being done in a way that at the same time improves certain key performance characteristics, like pulse-to-pulse stability. It also provides a unique flexibility in terms of pulsing and temporal pulse shaping.

It is simply not practical to make the individual laser oscillator larger and longer ad infinitum. Scaling issues like cooling, discharge uniformity, control of beam quality preclude this.

Excimer laser annealing system with 1300-mm line beam length

Each individual VYPER oscillator can deliver a total power of 1.2 W (2 joules) at a 6000-Hz repetition rate. The first four-oscillator system increased this to 2.4 kW in 2013, providing a line beam length of 1300 mm for large-format LTPS production on Gen 8 display glass panels. Meanwhile the latest three-laser system developed in 2016 provides up to 3.6 kW using the combined power of six oscillators.

The two oscillators of the VYPER are arranged in pairs with a master and slave configuration. Patented synchronization circuitry accurately controls the time delay between the pulses of the slave and master. An active feedback loop is employed to maintain a stable delay between the laser pulses over the running period, with a typical accuracy of 2 ns.

In this way, a two-oscillator system user can overlap the pulses to provide a high energy of 2 J per pulse at 600 Hz. Optionally, the pulse delay can be adjusted to provide longer pulses, with peak power sustained up to twice the normal laser pulse duration (24 ns). This allows exploration of the subtle effects of pulse duration on microcrystalline structure optimization. And, just as important, the concept of using multiple oscillators provides even further scalability of the line length for Gen 8+ glass panel generations.

Uncompromised line beam uniformity and pulse stability

There are two ways in which multiple excimer laser beams could be combined to make a single line profile. First, the laser beams can be combined with near 100

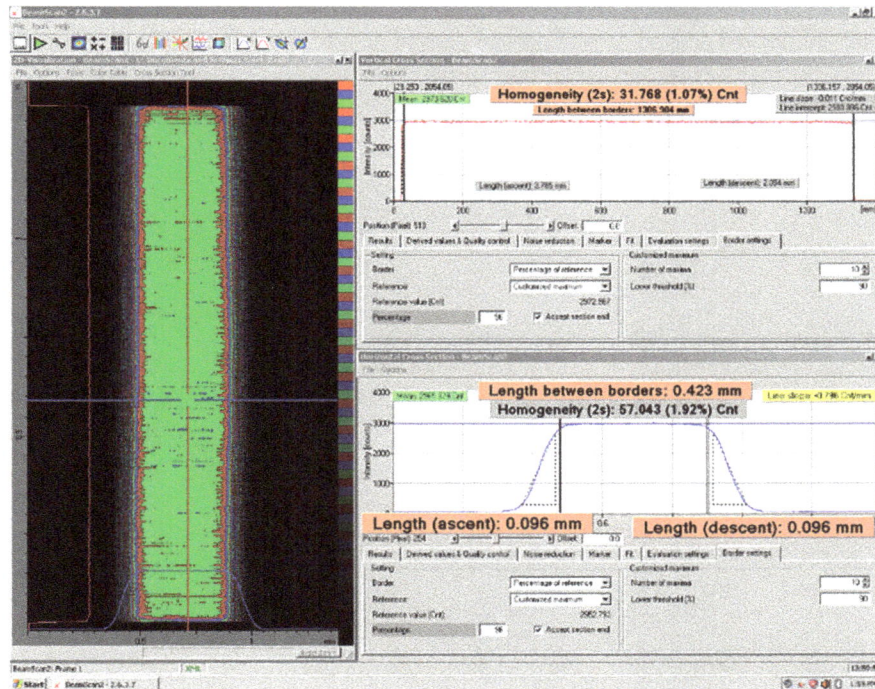

Anatomy of a 1300-mm process beam measured right in front of the display substrate.

percent overlap and the mixed beam then shaped to the final line profile. Or, the laser beams could be shaped, then stitched together end-to-end to form the final line. The former was chosen as this maximizes pulse-to-pulse stability at every location along the line.

Pulse-to-pulse stability is a particularly important process parameter impacting poly-Si crystal size uniformity. Specifically, the energy stability of each individual oscillator is between 0.25 and 0.3 percent sigma, well within the LTPS process window — in spite of the inevitable variations in repetitively firing a plasma discharge at tens of kilovolts. But, by combining and mixing the pulses from two oscillators, this random pulse-to-pulse noise is reduced by a factor of $\sqrt{2}$, for a final value of about 0.2 percent sigma.

The way in which the discharge is equilibrated between the pair of laser oscillators is another feature contributing to the low pulse-to-pulse variations and excellent timing synchronization of the two oscillators. The two laser tubes and their solid-state pulsed power supplies are, in principle, identical. The supplies are made identical in practice using a patented hardware module called EquiSwitch. This momentarily connects then disconnects the two-power storage systems so that they are equalized, immediately prior to firing the gas discharge. By eliminating any charge deviation on the storage capacitors, the switching behavior and the cycle time of energy through the circuitry is stabilized, resulting in long-term temporally synchronous output beams of nearly identical pulse energy.

Continuing evolution of laser annealing applications

In conclusion, excimer lasers have become essential to the fabrication of all advanced mobile displays we all now encounter on a daily basis. But, the technological needs of excimer laser annealing applications continue to rapidly evolve, concurrent with market pressure to reduce production costs. Excimer laser power and line length scaling have proven capable of meeting the market requirements, and straightforward system scale-up will carry this trend on into the future.

Meet the authors

Rainer Paetzel studied biomedical engineering at the University of Applied Sciences in Giessen, Germany. As director of strategic marketing for Coherent Laser Systems GmbH & Co. KG in Goettingen, Germany, his focus is on lasers systems used in flat panel display and microelectronics manufacturing. Ralph Delmdahl is product marketing manager for the Coherent Excimer Business Group. Delmdahl received his Ph.D. in physical chemistry from the Braunschweig University of Technology and his M.Sc. in economics from the Open University in Germany.

DLIP Quickly Changing Surface Functionalization

Direct laser interference patterning provides high fabrication speed and low fabrication costs.

BY ANDRÉS F. LASAGNI, TECHNISCHE UNIVERSITÄT DRESDEN
and FRAUNHOFER IWS

A wide range of products — from LEDs to solar cells to medical implants — can be improved through the use of micro- and nanostructures that significantly influence the physical properties of their surfaces. Such topographies can be utilized to functionalize surfaces and adjust their friction, wear, light management, biocompatibility or other properties to the specific requirements for a certain application. These micro- and nanostructures will have an impact on industrial applications only if they can be produced at high fabrication speed with low fabrication costs, and a new processing method called direct laser interference patterning (DLIP) meets both needs.

Periodic pattern formation

Traditionally, very few technologies have been capable of producing micro- and nanostructured surfaces on large areas. One is laser interference lithography (LIL). In LIL, the standing wave pattern that exists at the intersection of two or more laser beams is used to expose a photosensitive layer, also known as a resist. In the case of a negative resist, the positions that correspond to the interference

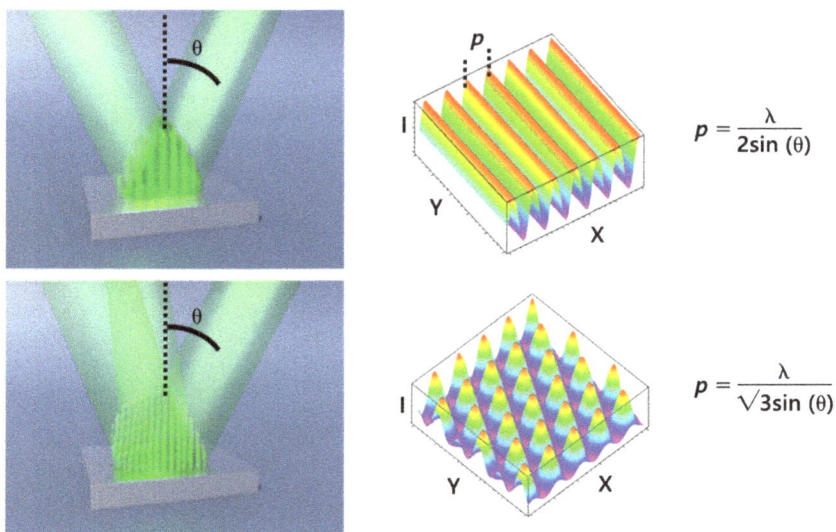

$$p = \frac{\lambda}{2\sin(\theta)}$$

$$p = \frac{\lambda}{\sqrt{3}\sin(\theta)}$$

Figure 1. Calculated intensity distribution for two-beam and three-beam interference patterning. The geometrical configurations of the beams necessary to achieve the displayed geometries are also shown. Courtesy of Fraunhofer IWS and Technische Universität Dresden.

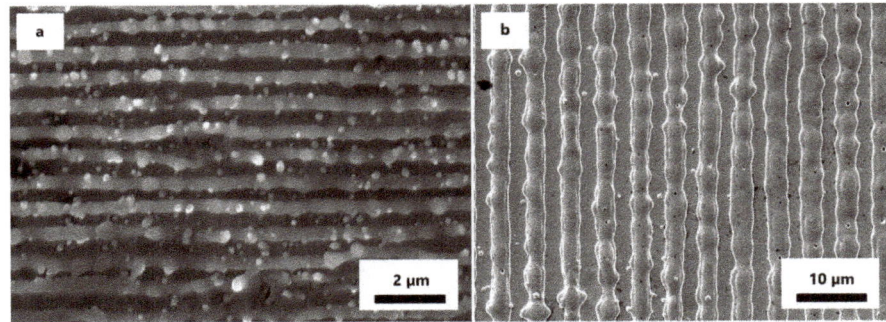

Figure 2. Structured stainless steel substrates using two-beam interference patterning using (a) picosecond and (b) nanosecond laser pulses. The spatial periods are 0.7 μm and 5.0 μm for (a) and (b), respectively. Courtesy of Fraunhofer IWS.

maxima are photopolymerized. After subsequent resist development, a periodic variation in the surface topography is obtained. However, the multistep character of LIL negatively influences the processing cost and only permits the processing of planar surfaces.

To overcome this problem, pulsed-laser systems with sufficient power and/or pulse energy can be used to directly ablate the surface of the material; hence the name direct laser interference patterning. Similarly to LIL, this technique enables the formation of periodic patterns with different features and defined long-range order. Depending on the number of laser beams that are used, and their geometrical arrangement, different surface geometries can be produced. A two-beam setup, for example, produces a line-like interference pattern, while three beams enable a dot-like geometry (Figure 1).

The most important requirement for the fabrication of periodic structures with the DLIP method is that the material being processed must be able to absorb the energy of the laser at the selected wavelength. If high-power laser systems are used, it is possible to achieve fabrication speeds of up to about 1 m^2/min. The structuring process is based on photothermal, photophysical or photochemical mechanisms, depending on the material type. In general, polymers and ceramics are processed with UV laser radiation, whereas green or IR lasers are applied to treat metals and coatings.

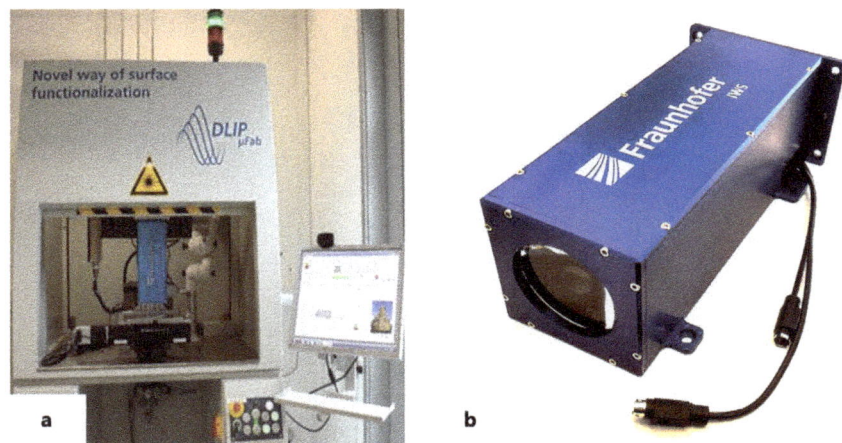

Figure 3. A direct laser interference patterning system (DLIP-μFAB) (a) and a DLIP optical head developed at Fraunhofer IWS and Technische Universität Dresden (b). Courtesy of Fraunhofer IWS and Technische Universität Dresden.

In the case of metals, DLIP is based on a photothermal process that involves local melting and/or selective ablation at the interference maxima positions. For nanosecond laser pulses, the primary material removal mechanism is ablation, but substantial melting occurs. The minimum pitch, or spatial period, is therefore limited by the thermal diffusion length, which is roughly 1 μm for stainless steel and titanium and 2 μm for copper (for 10-ns pulses). When the laser pulse width reduces to the order of picosecond or femtosecond, little thermal damage is observed and pitches below 1 μm are feasible (Figure 2).

At the laboratory scale, interference patterns can be obtained by splitting a coherent laser beam using beam splitters while the beams are later overlapped at the workpiece using several mirrors. Although these optical configurations are acceptable for research purposes, an industrial application of the DLIP technology will only be possible if compact optical-head solutions and systems are available.

Optical concepts and systems

In recent years, different laser interference patterning optical concepts and systems (DLIP-μFAB) have been developed (Figure 3). These systems offer the possibility of processing not only planar surfaces but also complex

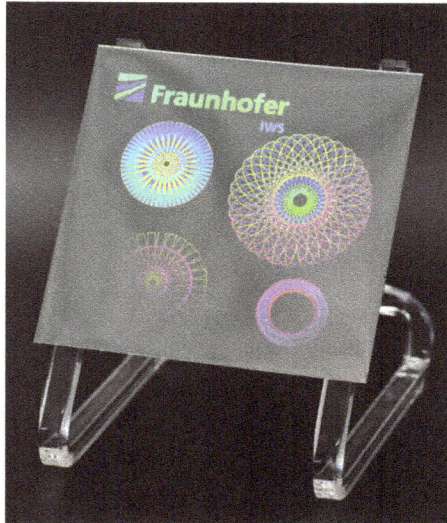

Figure 4. An optical photograph of a decorative element fabricated on nickel substrate using a nanosecond-pulsed IR laser system installed in the DLIP-μFAB machine. Courtesy of Fraunhofer IWS.

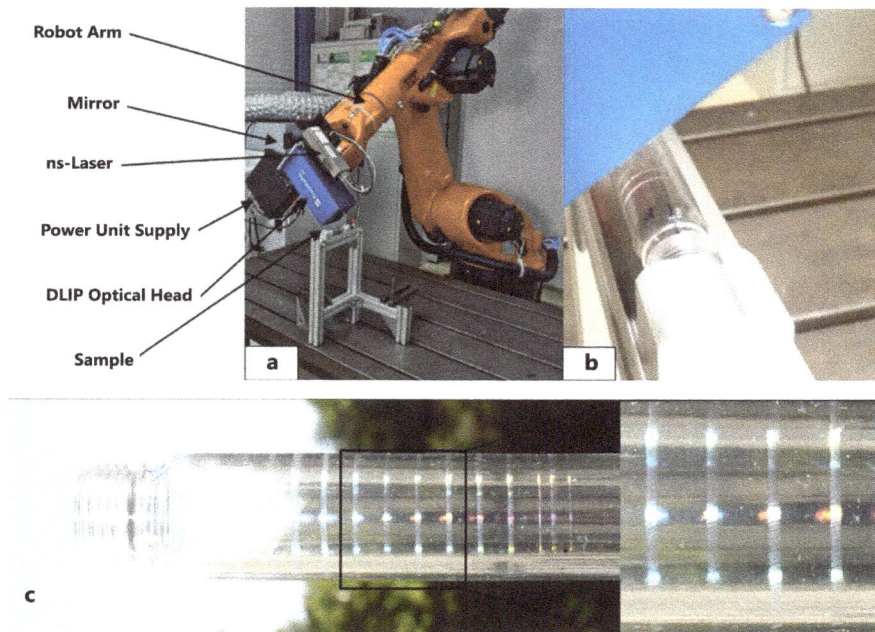

Figure 5. The experimental setup of DLIP using a robot arm. The DLIP optical head is fixed to the robot arm together with the laser system as well as the power unit supply (a). An optical micrograph of the cylindrical polyethylene terephthalate (PET)-treated part and the used DLIP optical head over the cylinder (b). A treated PET cylinder at different positions corresponding to different linear speeds. The inset shows the structures produced at 400 mm/s with a positioning error of ~90 μm (c). Courtesy of Fraunhofer IWS.

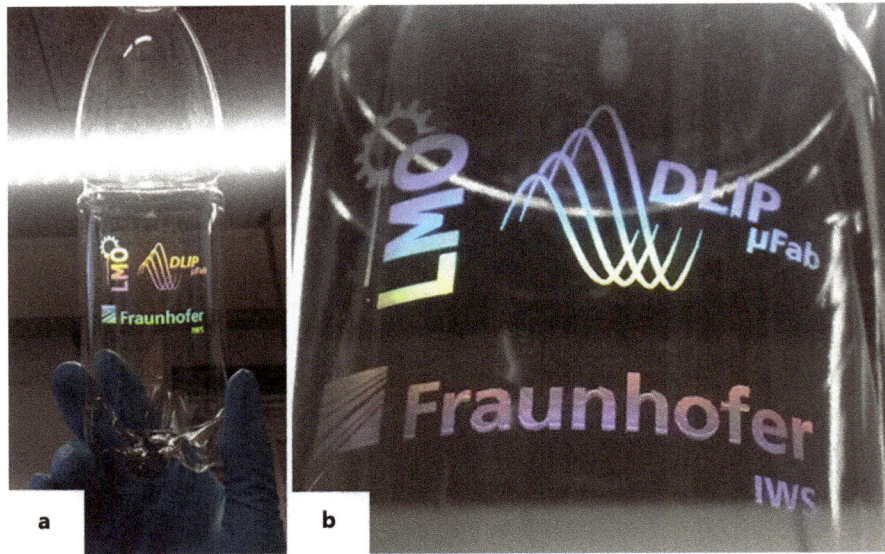

Figure 6. A treated PET bottle using a DLIP-µFAB system (a) and (b). The accuracy of the system is below 1 µm. The system is equipped with 4 axes, including 200 × 200-mm XY translational states and a 360° rotational axis. The DLIP-optical head is mounted over a Z-translational stage with travel length of 100 mm. Courtesy of Technische Universität Dresden.

3D parts. The optical heads (Figure 3b) can be equipped with different components that control the intercepting angle between two laser beams fully automatically. Consequently, manufacturerscan fabricate periodic structures with different spatial periods at different positions. In this way, decorative elements, for instance, can be produced (Figure 4).

In addition to 2D parts, DLIP can be used for the treatment of 3D elements. For this purpose, the DLIP optical heads can be combined with either linear or rotational translation stages or they can be mounted on a robotic arm.

3D processing

When combined with linear or rotational translation stages, the DLIP processing head can be translated vertically, and the relative Z-position of the two components can be adjusted. This strategy is appropriate for parts with curvature angles that are less than 20°, corresponding to deviations in the structure period of up to 6.4 percent. If it is necessary to treat 3D parts with larger angles of curvature, an industrial robot can provide the required flexibility (Figure 5a). The main advantages of this system are the large working space — 2033 mm, 2230 mm and 2429 mm — and a maximum linear speed of 2000 mm/s. A 263-nm laser system would

DLIP Highlights

- Capable of treating large surface areas
- Processes micro- and nanostructures on planar surfaces
- Processes micro- and nanostructures on complex 3D parts
- Fabrication speeds up to 1 m²/min (with high-power lasers)
- Submicrometer resolution

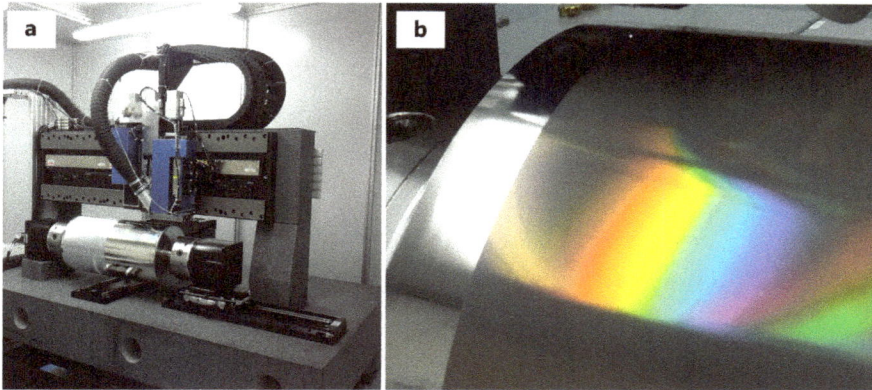

Figure 7. The large area DLIP system for the processing of 2D and 3D parts (a). A 300-mm-diameter nickel sleeve processed with a nanosecond-DLIP optical head (b). Courtesy of Technische Universität Dresden.

be used, for example, because of the high absorption of polyethylene terephthalate (PET) bottles in the UV range (Figures 5b and 5c).

If rotational symmetrical parts have to be treated, the DLIP-µFAB system can be equipped with a rotational axis. With the processing of a PET bottle, for example, it must be fixed to the rotational axis and be translated with a linear stage and rotated to produce 3D elements, such as holographic pixels, directly on the object's surface (Figure 6). Due to the high accuracy of the used translational stages and rotation axis, the accuracy between the treated area per laser pulse was approximately 0.5 µm.

Larger parts

When larger parts must be treated, a special DLIP system, developed at Germany's Technische Universität Dresden, can be used (Figure 7). This system treats cylindrical parts up to 600 mm in length and 300 mm in diameter, and is equipped with both nanosecond and picosecond laser systems. Similar to the DLIP-µFAB system, different optical heads can be utilized.

These results demonstrate the capabilities of the DLIP method for the production of periodic surface patterns on 2D and 3D parts, with resolutions down to the submicrometer range. Further, these results indicate the high maturity level of the technology and highlight its potential for integration into industrial applications.

Meet the author

Andrés F. Lasagni is the group leader at the Fraunhofer IWS and a professor at the Technische Universität Dresden, both in Dresden, Germany.

Laser Paint Removal Takes Off in Aerospace

Environmentally harmful paint-stripping methods are being replaced with laser ablation.

BY JAMES SCHLETT, CONTRIBUTING EDITOR

At 52 feet tall, over 100,000 pounds and with a reach of 85 feet, LR Systems' laser ablation mobile robot is one of the largest — if not the largest — mobile robots in the world. This summer, the company, based in the Netherlands, plans to install a laser coating removal robot (LCR) at the maintenance facility of a yet-to-be-announced large airline (Figure 1). The LCR, which features a newly designed 20-kW CO_2 laser, will be the first full-aircraft, robotic laser stripping solution to be deployed by a commercial air carrier. The robot's deployment will represent a milestone in the decades-long effort, largely driven by the U.S. military, to replace

Figure 1. A rendering of LR Systems' largest laser coating removal (LCR) system for commercial airliners. Courtesy of LR Systems.

costly and environmentally harmful aerospace paint-stripping methods, such as solvents, abrasives and blasting, with laser ablation. When mounted to the robot, the laser will be able to deliver a precisely aligned beam over 30 meters to remove microns of paint.

"It is really many factors coming together, including advancements in robotics, lower cost and higher performing lasers, a greater emphasis on the sustainability and health benefits of the process, and private investment," said Clay Flannigan, the assistant director of robotics and automation engineering at Southwest Research Institute (SwRI) in San Antonio. "To date, the U.S. Department of Defense has funded most of the process development and we are just now starting to see the commercialization of the technologies."

Despite its Dutch roots, the LCR is the fruit of a collaborative with partners on both sides of the Atlantic. SwRI is the project's prime contractor for the robot's development, including a high-speed machine vision system for real-time control of laser power for selective stripping of paint layers. The Edison Welding Institute in Columbus, Ohio, provided a polygon laser scanner, which will focus and sweep the laser across the aircraft's surface. The CO_2 laser was developed by Trumpf Nederland in the Netherlands. LR Systems also has partnered with the Netherlands' National Aerospace Laboratory and the Dutch Aerospace Lab for assistance in the qualification process with the U.S. Federal Aviation Administration and the European Aviation Safety Agency.

Figure 2. General Lasertronics uses an Nd:YAG laser to remove coatings from CH-53 heavy-lift transport helicopters. Courtesy of Sikorsky.

LR Systems' website identifies Singapore Airlines as a first potential customer. The airline's subsidiary, the Singapore Airlines Engineering Co., has entered into a memorandum of understanding with the Dutch company to provide support through the OEM qualification and certification process.

De-painting basics

Along with adding color to an aircraft, paint adds weight. For example, Airbus' double-deck, wide-body, four-engine jet airliner A380, which features a 4,400-square-meter exterior surface, requires three layers of paint that weigh a total of approximately 500 kg[1]. Paint can either enhance marketing for commercial airlines through the use of company colors and logos or create strategic advantage for military jets in the form of camouflage designs. Simply painting on top of an old layer of paint that has begun to show wear could make the plane look new but at the expense of added weight.

'Transitioning laser ablation from laboratory coupon scale to manufacturing scale is the focus of ongoing research.'

— *Kay Blohowiak, a senior technical fellow in chemical technologies and adhesive bonding at Boeing.*

Vision Systems Send Paint Stripping Into High Speed

Needing to balance aesthetics with weight, aerospace manufacturers typically apply thin layers of paint to their aircraft. While planes coming out of the factory may have up to 5 mils of paint, that thickness can be up to 20 mils for aircraft that have long been in operation and subjected to touch-ups, according to Jim Russell, director of business development for General Lasertronics, a San Jose, Calif., defense contractor. To precisely strip these thin layers of paint from aircraft — sometimes stopping short of the primer and without damaging the underlying aluminum or composite material — robotic laser coating removal systems rely on a variety of optical and laser sensors.

However, not all robots see laser paint removal the same way because of the different high-speed vision systems they employ. Below are brief descriptions of the vision systems used in the Laser Coating Removal System (LCR) by LR Systems, the Automated Rotor Blade Stripping System (ARBSS) by General Lasertronics and the Advanced Robotic Laser Coating Removal System (ARLCRS) by Concurrent Technologies Corp. and the National Robotics Engineering Center.

LCR: The system monitors the estimated state of paint removal by using color and NIR imaging at a rate of 300 Hz. The vision system provides a laser power command with a resolution of several millimeters. Laser power is updated 20,000 times a second. The LCR "learns" what the desired paint looks like by using machine learning algorithms, according to Clay Flannigan, the assistant director of robotics and automation engineering at San Antonio-based Southwest Research Institute, which developed the LCR's vision system.

ARBSS: The system evaluates the color of the work surface, with laser pulses being held in check until the presence of the coating to be removed is confirmed. This system selectively removes a top or sub-coat of paint and prevents the laser from ever directly striking the delicate fiberglass composite substrate, according to Russell.

ARLCRS: The system's color cameras provide data for a surface property analyzer and a scanning lidar sensor collects 3D data for surface mapping[1]. The sensors detect and classify the state of de-painting, allowing the robot to modify its coverage plans and avoid stripped areas or target areas where coatings remain[2].

To learn more about high-speed vision systems, see the April 2014 issue of *Industrial Photonics* for a technical feature article, "Vision Offers Control of a High-Power Laser Ablation Process," by Michael Rigney, a staff engineer in SwRI's Robotics and Automation Engineering Section. Go to www.photonics.com/A56086.

References

1. AFSC/LGMI (2014). Air Force full-aircraft robotic laser paint removal. Video. https://www.youtube.com/watch?v=MEbxPi8uo78.
2. Carnegie Melon University. CMU and CTC to develop robotic laser system to strip paint from aircraft. Press release, Nov. 26, 2012. http://www.eurekalert.org/pub_releases/2012-11/cmu-cac112612.php.

"Each five or six years an aircraft is stripped and painted again to support repair of damages on the surface … and redo the paint to shield the aircraft from corrosion as paint needs to be replaced," said Peter Boeijink, LR Systems' program director.

Traditionally, de-painting has been done with chemical paint strippers, water picks, dry media blasting and hand sanding. In 2007, for example, to strip large parts of Boeing KC-135 air tankers that are usually removed during depot maintenance, U.S. Air Force personnel at Tinker Air Force Base in Oklahoma City used 4,360 gallons of chemical paint removal and 2.7 million gallons of rinse water that became contaminated during the de-painting process[2]. Since the early 1990s, the Air Force has been exploring alternative de-painting methods[3]. That search became more pressing with the enactment of federal legislation such as the Clean Water Act, Clean Air Act and Resource Conservation and Recovery Act, which put restrictions on the discharging of wastewater with hazardous waste, the emission of hazardous air pollutants and the disposal of de-painting-related waste, respectively.

For every one pound of de-paint, about four pounds of solid waste are produced through media blasting, and nine pounds of solid waste — plus 165 pounds of liquid waste — are created through the use of solvents. But with laser paint stripping, only about a half pound of solid waste results, according to defense contractor General Lasertronics, of San Jose, Calif., whose installed laser-based Automated Rotor Blade Stripping System (ARBSS) is used to remove coatings from Navy helicopter rotor blades (Figure 2).

Laser demand

When Lasertronics launched its first paint stripping system in the mid-1990s, it did so with a CO_2 laser. The company used this technology until about 2000, when it was replaced with an Nd:YAG laser from Northrop Grumman's Cutting

Edge Optronics in St. Charles, Mo. Jim Russell, Lasertronics' director of business development, said the Nd:YAG's flexible fiber optic beam delivery system enables "access to nooks and crannies on aircraft," such as internal wing fuel tanks. Although the company's customers were pleased with the Nd:YAG, they were less so with the downtime the technology required and its frequent need for costly maintenance. Consequently, Lasertronics in 2014 started using fiber lasers from IPG Photonics in Oxford, Mass.

"The IPG fiber laser has proven to be extremely robust and operator-friendly. As laser prices continue to fall, we envision many coating removal jobs, across industries, being taken over by laser ablation systems," said Russell.

The bubbling aerospace laser ablation market has not gone unnoticed at IPG. In an April 2015 conference call, IPG Chairman and CEO Valentin Gapontsev noted the receipt of "high volume requests for efficient paint removal systems for aircraft and ships." He added that the company's long-term project for this application was "turning from qualification to the mass deployment phase, and IPG fiber lasers are playing the role of an excellent engine there." Last February he said during a conference call that laser cleaning, including in aerospace, was a growth area for IPG. He added that the company's "unique multi-kilowatt nanosecond fiber lasers are starting to change the situation in this large market segment."

Lasertronics was not alone in transitioning away from CO_2 lasers for robotic paint removal systems. Concurrent Technologies Corp., which is based in Johnstown, Pa.,. also utilizes IPG fiber lasers for its Advanced Robotic Laser Coating Removal Systems (ARLCRSs), which are used by the U.S. Air Force. LR Systems, however, still uses CO_2 lasers. In fact, its CO_2 laser by Trumpf is believed to be the largest of its kind for laser ablation for its robotic system. Flannigan at the Southwest Research Institute said this laser is especially good at removing difficult coatings, such as clear coats and the white top coats that are prevalent on commercial airliners.

"White paints tend to be highly reflective and low solids paints, such as clear coats and primers, tend to be highly transmissive. Both of these phenomena have [a] negative impact on stripping efficiency and substrate temperatures. CO_2 wavelengths, 10.6 microns, are well-absorbed by most paints, and thus, more energy goes into ablating the paint than heating the substrate or environment," Flannigan said.

While pulsed lasers' high peak power help them overcome the limitations imposed by continuous lasers' shorter wavelength, Flannigan said the paint-stripping rate is approximately proportional to average laser power. That makes it difficult to safely deliver a high average power pulsed laser without locally overheating the substrate.

"The issue with the CO_2 lasers is that the beam needs to be manipulated around the aircraft using mirrors, and those mirrors need to maintain alignment with each other — a very difficult task. In addition, the cost of operating and maintaining a fiber is significantly less than a CO_2 system," said Tom Nugay, deputy technical director for the Air Force Materiel Command's Logistics, Civil Engineering and Force Protection Directorate at Wright-Patterson Air Force Base in Ohio.

Military advances: jets

Recent years have seen laser de-painting make significant gains at U.S. air bases, particularly since the 2008 demonstration of a Robotic Laser Coating Removal System (RLCRS) on KC-135 parts at Tinker Air Force Base. The RLCRS was the one of a series of robotic systems developed by Concurrent for the Air Force, and a year later the company built and installed a Laser Automated De-painting System II (LADS II) at Hill Air Force Base in Utah. The Air Force used this railed robotic system with a Rofin-Sinar Technologies CO_2 laser to strip General Dynamics' F-16 fighter jets.

In June 2015, Concurrent and Carnegie Mellon University's National Robotics Engineering Center in Pittsburgh developed and transferred to the Utah air base two ARLCRSs (Figure 3). Unlike the LADS II, the ARLCRSs are semi-autonomous and each robot features a fiber laser by IPG. Between the two systems, there are six robots, said Concurrent Principal Engineer Mary Bush, with two reserved for stripping F-16s and four for Lockheed Martin C-130 cargo planes. The ARLCRSs have reduced de-paint flow days by over 50 percent: from seven to three days for the F-16s and from 10 to five days for the C-130s, according to Nugay.

"The ARLCRS ... is the only semi-autonomous robotic laser system that has been developed, constructed and implemented. The technology is applicable to commercial airlines and other aerospace industries," Bush said.

ARLCRS regular production recently began on Hill Air Force Base's F-16s and is expected to begin on the base's C-130s during the first quarter of 2017 (Figure 4). The Air Force also plans to build next-generation ARLCRSs at Robins Air Force base in Georgia during the 2017 fiscal year. Nugay said that depending on the size of the aircraft, the systems cost between $5 million and $7 million.

"We believe that laser technology is very attractive for removing paint from composites, aluminum and steel materials used in the aerospace industry." Nugay said. "In addition, for each F-16, we see a 99 percent reduction in total waste disposal and an elimination of 2,000 pounds of hazardous waste, no hazardous materials used, minimal personal protective equipment and a clean control room working environment. That is why we are pursuing [it]."

Military advances: helicopters

In 2009, Lasertronics also installed its Automated Rotor Blade Stripping Systems at Marine Corps Air Station Cherry Point in North Carolina, where it strips paint off the fiber glass composite rotors of Sikorsky CH-53E heavy-lift helicopters. The laser system has eliminated the amount of scrap from parts damaged during de-painting, which had run above 10 percent with manual rotary sanders — an important reduction given that each blade is about 38 feet long and costs about $120,000.

"This is a nondestructive technology," said Ralph Miller, a Lasertronics co-founder and vice president of marketing communications.

Lasertronics is replacing the ARBSS's Nd:YAG lasers by U.S. Laser Corp. of

Figure 4. An Advanced Robotic Laser Coating System removes coatings to the primer layer of an F-16 at Hill Air Force Base, Utah. Courtesy of U.S. Air Force, photos by Alex R. Lloyd.

Wycoff, N.J., with 400-W ones made by Lee Laser Inc. in Orlando, Fla. After that upgrade is complete, Naval Air Systems Command plans extend the ARBSS program to the Boeing V-22 Osprey. Cherry Point's Sikorsky H-60 Blackhawk blades are also being eyed for laser stripping, said Lasertonics' Russell.

Rotor blades routinely need to be stripped both for blade repair and because touch-up paint unevenly applied in the field could result in balance problems, said Russell. Lasertronics is also exploring removing contact erosion guard coating from Boeing Apache rotor blades. A Lasertronics handheld laser system has also been used to remove sealant from a Fairchild Republic A-10 Thunderbolt's center wing fuel tanks, where it demonstrated an ability to dramatically throughput time and cost.

On the horizon: commercial airliners

Despite these inroads, Russell said laser stripping's growth in the military has been stunted by the automatic federal budget cuts, referred to as sequestration, first implemented on the U.S. Department of Defense in 2013. For Lasertronics, progress with commercial airlines has also been much slower than expected, especially after the company became the only laser de-painting company to receive FAA approval in 2010. While that agency's blessing is important, Russell said maintenance, repair and overhaul providers generally have been reluctant to adopt laser stripping technology because it is not included in OEM service manuals.

"As soon as the manuals can get changed, a whole lot of things will happen quickly," said Russell.

Commercial airliner interest

That is not to say commercial airliners have been reluctant to embrace laser ablation. Since the early 2000s, for example, Airbus has used lasers made by Adapt Laser Systems of Kansas City, Mo., to remove paint for electrical bonding contacts. Adapt already has approximately 30 lasers in use for aircraft paint removal tasks, including several fiber lasers for small-area paint removal on the F-16s and C-130s at Hill Air Force Base. In 2015, Lockheed Martin also approved an Adapt laser for the removal of coatings from areas for structural bonding of nutplates to metallic structures of F-35 multirole stealth fighters, according to Adapt Laser President Georg Heidelmann. He is "very optimistic" that lasers will be deployed in the near future for full-body de-painting of military and commercial aircraft.

At Boeing, Kay Blohowiak, a senior technical fellow in chemical technologies and adhesive bonding said the company "... continues to research new technologies to improve our manufacturing processes, including the use of lasers to replace chemicals, and manual sanding processes for removing coatings from aircraft structures, as well as for cleaning and preparing surfaces for painting, bonding and sealing. Transitioning laser ablation from laboratory coupon scale to manufacturing scale is the focus of ongoing research."

References

1. British Airways, Fleet Facts: Airbus 380-800. http://www.britishairways.com/en-gb/information/about-ba/fleet-facts/airbus-380-800.
2. J. Arthur, et al. (2008). Remote Laser Coating Removal System. Final Technical Report. http://www.dtic.mil/dtic/tr/fulltext/u2/a608206.pdf.
3. Department of Defense Office of the Inspector General (April 1993). Air Force Study on Paint Stripping Technology. http://www.dodig.mil/Audit/Audit2/93-086.pdf.

Additive Manufacturing

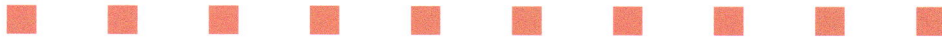

■ ■ ■ ■ ■ ■ ■ ■ ■ ■

Ultrafast Fiber Laser Opens Doors for Additive Manufacturing

By taking advantage of instantaneous high-temperature plasma generation, high-temperature metals such as tungsten can be transformed.

BY JIAN LIU, POLARONYX INC.

Femtosecond (fs) lasers have long been a workhorse in subtractive manufacturing, prized for their unique ability to athermally ablate materials. They are commonly used in surface structuring, drilling and thin-film scribing. However, few thought that an fs laser could be used in additive manufacturing (AM). By taking advantage of instantaneous high-temperature plasma generation, a recently developed fs fiber laser can melt high-temperature metals such as tungsten. Employing the fs fiber laser, parts created using tungsten achieved 99 percent density. Moreover, researchers have shown that the fs laser can deposit metals on glass substrate.

Overcoming limitations

Laser additive manufacturing centers on selective laser melting, using material powders to build three-dimensional parts with complicated structures. It's an efficient, robust and cost-effective technique for the next generation of manufacturing.

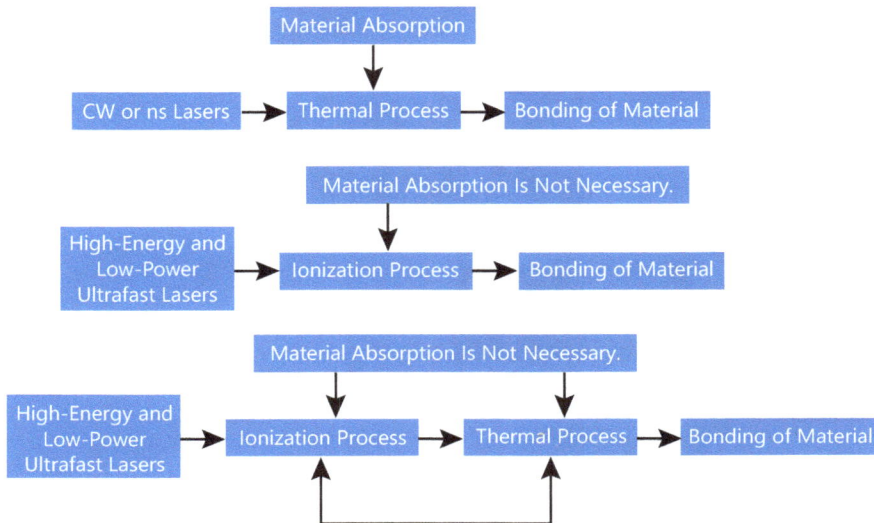

Figure 1. Comparison of femtosecond laser additive manufacturing process versus CW or ns laser process.

AM processes for many industrial metals such as titanium (Ti), with ~1668 °C melting temperature, and aluminum (Al), with ~650 °C melting temperature, are well-established. Melting these metals requires the use of a CW or long-pulsed laser.

However, when it comes to using current AM lasers and tungsten as the material of choice, challenges emerge, including residual stresses, density, uniformity and variation in mechanical strengths.

Notably, CW or long-pulsed lasers can only process materials with low- to medium-thermal conductivity and melting temperature, due to low peak intensity. Processing high-temperature and high-thermal conductivity materials such as tungsten — with a melting temperature of 3422 °C and thermal conductivity of 173 W/(m·K) — needs a much higher power laser to deposit high energy in a short period of time against fast thermal drain.

Another problem is that simply increasing the power of a CW or long-pulsed laser induces thermal diffusion outside the focal volume — often referred to as HAZ — as well as residual stress. That can cause cracks and fatigue of the joint part. Furthermore, the formation of unstable intermetallic phases can further degrade the quality of joining and reduce hardness and strength. Separate post-processes are also required to polish, cut, trim and structure the AM components, leading to extra labor and cost.

Finally, current CW laser additive manufacturing systems use material absorption to bond powders — but for bonding to occur, the materials can only absorb at a specific laser wavelength.

Recently, a team of researchers developed an fs laser for AM for melting and shaping tungsten powders, hafnium diboride (HfB_2) and zirconium diboride

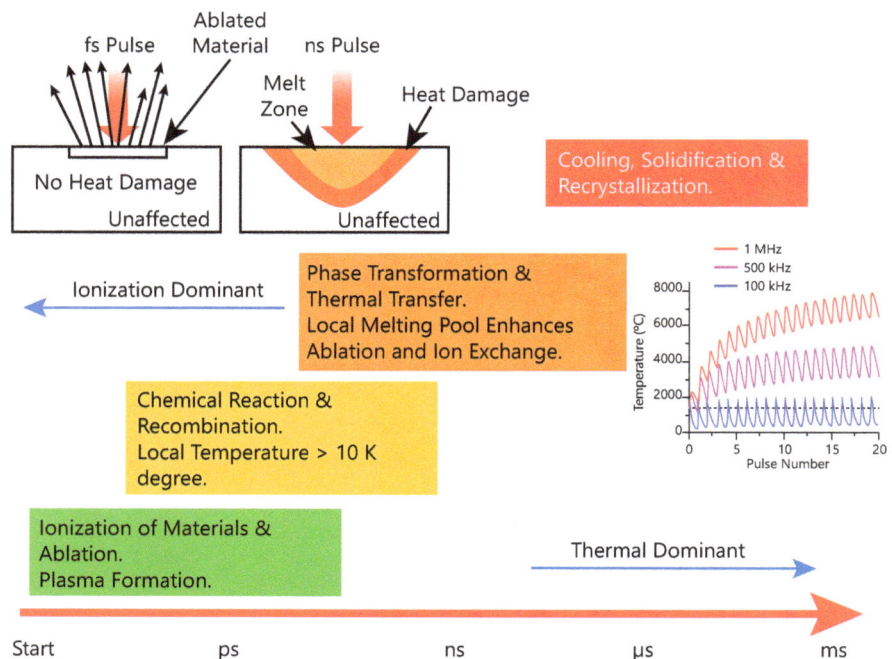

Figure 2. Temporal evolution of fs laser additive manufacturing process.

(ZrB_2), providing an unprecedented way to modify material functions and mechanical properties[1,2,3,4].

Rapid delivery of energy

The main characteristic of the ultrashort laser pulse is the high-peak intensity that brings about the rapid (ps) delivery of energy into the material, independent of material absorption characteristics, to cause ionization, which is much faster than the plasma expansion (ns to μs). As a result, the local temperature is rapidly increased to over 6000 °C (controllable through energy and pulse number) and the thermal damages to surroundings are reduced or eliminated.

Compared with CW laser additive manufacturing, the fs laser approach creates instantaneous high temperatures to melt high-temperature and high-thermal-conductivity metals, forming much stronger microscale welding/bonding between similar or dissimilar refractory metal powders in various shapes and sizes. This multifunctionality could significantly reduce building time and cost, which is not achievable for CW laser additive manufacturing. A comparison of the mechanism using CW, high-energy low-pulse repetition rate and high-energy high-pulse repetition rate highlights the benefit of balance of ionization and thermal process (Figure 1).

Many parameters can have an impact on fs laser additive manufacturing quality. The process is very complicated, involving variables related to material properties, laser parameters, as well as human operation and intervention. In terms of laser parameters, there are energy, pulse width, average power, pulse repetition rate, peak power, beam quality, focal spot size, hatch, scanning speed and contour, and mode of operation. In terms of powder quality — size, shape and residual stress all play critical roles.

As to the dynamics of powder welding — heat flow, chemical reaction, metal evaporation, thermal diffusion and transfer, stress and fatigue are important. Melting, solidification and cooling, grain/microstructure formation, phase transformation, cracking, and femtochemistry are important parameters, too.

	Iron	Glass
Melting Temperature (C)	1400	1500
Thermal Conductivity W/(m·K)	17	0.9
Thermal Expansion Coefficient (μm/m·K)	15.9	5.9
Thermal Diffusivity (mm^2/s)	3.3	0.34

Table 1. Comparison of iron and glass.

Closer look at the AM process

A simplified look at the mechanisms of the AM process (Figure 2) shows that the ablation of the fs fiber laser incurs ionization and recombination of materials in the ps regime to form new grains and microstructures during supercooling and solidification from a few ps to a few ms.

By varying process parameters such as fs fiber laser parameters (energy, power, pulse repetition rate), scanning speed and pattern, any type of sample can be made with controllable porosity, microstructure, density (up to 99 percent), shapes and structures.

Tungsten powders can be used to make various components, including thin walls and gears, on tungsten substrates (Figure 3). It is also important to note that the microstructure can be varied by changing the pulse width (Figure 4) to tailor the mechanical properties of AM parts[3].

Fabricating dissimilar materials

Moreover, femtosecond laser additive manufacturing shows unprecedented advantages in dissimilar material parts fabrication. As an example, iron was chosen as the main material for investigation; it has a melting temperature of ~1400 °C and its heat conductivity at room temperature is 17 W/(m·K)$^{-1}$. The iron powders have a size distribution of 1 to 5 μm. Three-mm-thick glass slides were used as the substrates.

Iron and glass feature starkly different melting temperatures and thermal conductivity (Table 1). In addition to those differences, glass is very brittle and easily cracked during the laser process.

A side coating of a fused silica glass window was created to reduce stray light

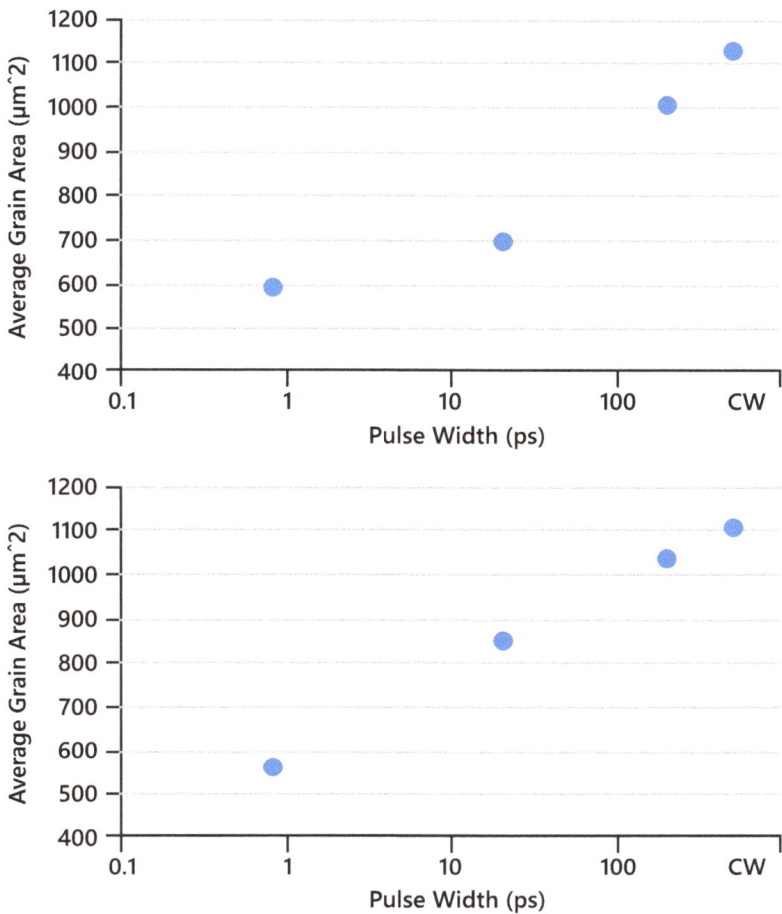

Figure 4. Average grain size as a function of pulse width. Top surface (top); cross section (bottom).

for laser remote sensing. Three layers of iron powder were spread and melted for all four sides of the glass window (Figure 5). The reflectance (<1 percent) improvement is significant over uncoated surfaces of the window. After three layers of iron powder coating, the surface of the glass was completely covered with melted iron powder (Figure 5a), which completely blocked the stray light when a red light from a laser pointer passed through the glass windows (Figure 5b).

Figure 5. Laser light passes through an optical glass window. An iron powder coating is applied to four sides of a fused glass window, helping minimize stray light (a), compared to laser light passing through an uncoated optical glass window (b).

Applications in developing new medical devices

Femtosecond fiber laser AM shows unprecedented capability in melting refractory materials, making multimaterial components, and manipulating microstructures to tune mechanical properties. By integrating the ablation feature of a femtosecond laser, the 3D printing system can make devices with more complex structure. These features enable many applications in aerospace, medical devices and defense.

Meet the author

Jian Liu is the founder and president of PolarOnyx Inc. He has pioneered and led the product development of a femtosecond laser 3D printing system (Tungsten-LAM, 2016 R&D 100 Award winner).

References

1. Bai Nie, et al. (2015). Femtosecond laser additive manufacturing of iron and tungsten parts. *Appl Phys A*, DOI: 10.1007/s00339-015-9070-y.
2. Bai Nie, et al. (2015). Femtosecond laser melting and resolidifying of high-temperature powder materials. *Appl Phys A*, Vol. 118, Issue No. 1, pp. 37-41.
3. S. Bai, et al. (2016). Manipulation of microstructure in laser additive manufacturing. *Appl Phys A*, Vol. 122, p. 495, DOI 10.1007/s00339-016-0023-x.
4. S. Bai and J. Liu, et al. (February 15-18, 2016). Femtosecond fiber laser additive manufacturing of tungsten. SPIE 9738-24 (Invited Talk), Photonics West 2016, San Francisco.
5. J. Liu, et al. (2016). Glass surface metal deposition with high power femtosecond fiber laser. Submitted to *Appl Phys A*.

Additive Manufacturing: The Laser Source Is Critical

Laser beam power, spot size and focal point must be measured and controlled to produce parts to specification through 3D printing.

BY DICK RIELEY, OPHIR-SPIRICON LLC

Laser-based additive manufacturing (AM) involves the deposition of a precise amount of material in a layer-by-layer process to produce components of specific measurements and physical properties. Instead of milling a workpiece from a solid block, AM builds up components layer by layer using materials that are available in a powder form. A wide variety of materials from plastics to metals and even composite materials can be used.

The process is dependent upon the precise delivery of the laser heat source to produce the deposition, usually by melting a powder feedstock in a predictable thickness, layer upon layer, in a fully fused manner. Although the term "3D printing" is on the rise, additive manufacturing is a more accurate description, because it captures the method's distinct difference from conventional approaches to materials removal.

Several variables impact the quality of an additive manufacturing process. The laser source is critical, especially in terms of its power, spot size and focal point. If any of these three variables is not properly measured and controlled, the likelihood of producing a part to specification is improbable.

In a recent application with the Center for Innovative Materials Processing through Direct Digital Deposition (CIMP-3D) at Pennsylvania State University in State College, Pa., a complete beam-diagnostics procedure was conducted on one of the lasers used for additive manufacturing research. The system evaluated was a 500-W fiber laser from IPG Photonics with a 50-µm core fiber. It was fitted with a custom lens assembly that could deliver an estimated 50-µm focal spot to the point of processing, enabling very fine features to be fabricated during deposition.

The first part of the test involved verifying that the selected power level entered by the operator and controlled by the laser power supply was indeed delivered to the work surface. If the amount of power requested is not what is delivered, the process may not perform as desired, and there is a potential for a lack of fusion of the AM powder materials, which could lead to defects in the component under construction.

Setup: Power measurement

To conduct this test, the laser optical head was positioned so that the beam diameter on the Ophir 500-W fan-cooled thermal sensor would remain below the

> Although the term '3D printing' is on the rise, additive manufacturing is a more accurate description, because it captures the method's distinct difference from conventional approaches to materials removal.

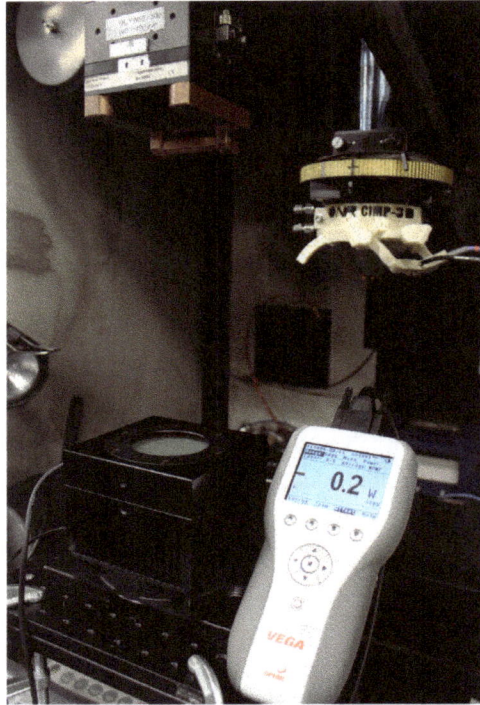

Figure 1. This experimental setup to measure laser beam power includes an optical head with copper collar, a 500-W fan-cooled sensor and the Ophir Vega digital display meter. All images courtesy of Ophir-Spiricon LLC.

damage threshold of the absorber disk. With a beam spot size of 5 mm, the power density (irradiance) of the beam was estimated to be 2546 W/cm^2, which corresponds to 33 percent of the damage threshold of the sensor, a very safe level.

A series of tests was conducted involving increasing laser power in 100-W increments — 100, 200 and up to 500 W. At each step, the operator recorded the input power, measured the corresponding power setting on the laser controller, and took a final measurement of the actual power output on the 500-W sensor at the work surface of the laser.

Figure 1 shows the experimental setup. The optical head is visible with the copper collar, the 500-W fan-cooled sensor is located directly below the optical sensor, and the Ophir Vega digital display meter (used as a readout device for each measurement) is directly to the right of the sensor. In a production environment, the Ophir Juno would be used in place of the Vega meter, because the Juno allows the sensor to be connected directly to the PC, providing a live feed of the data to the monitoring controller or PC.

Results: Beam power

Figure 2 shows a plot of the results. The first measurement showed that, as expected, the actual power delivered to the part is lower than the set point. However, it is interesting to note that the discrepancy is a significantly larger percentage of full scale at low power compared with higher power settings. In the chart, the blue line represents the set point at each level at which the laser should be performing, the red line is the display output on the laser controller, and the green line is the actual delivered power to the 500-W sensor at the work surface.

The initial summary of this data is that when the laser is operated at a lower power setting, a discrepancy of as much as 29 percent can exist between the set value and the actual delivered power. In comparison, when the laser setting calls for a full 500 W, the actual delivered power is only 11 percent less. This test provides a performance baseline for this laser. When subsequent power tests are conducted, the results can easily determine stability or change from this baseline. If there is a change, the appropriate diagnosis should be made and corrective action taken.

This information also affords the operator the opportunity to determine within the laser what is causing the discrepancy between input setting and output results. With each corrective action, a follow power can then be conducted to validate the action as correct or ineffectual.

Beam size, location

The second part of the experiment involves measuring the beam size and focal-point position. To accomplish this, the Ophir Photon NanoScan rotating slit detector was used with a 9 mm × 5 µm slit. In this setup, the optical head of the laser was positioned to deliver the estimated 50-µm beam directly onto the scan-

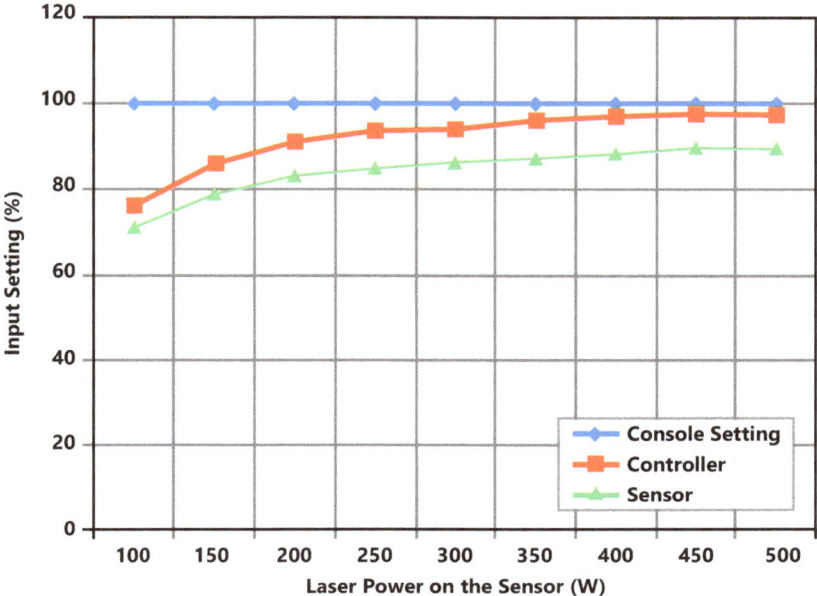

Figure 2. Plot of laser beam measurement results. The actual power delivered to the part is lower than the set point, but the discrepancy is a significantly larger percentage of full scale at low power compared with higher-power settings.

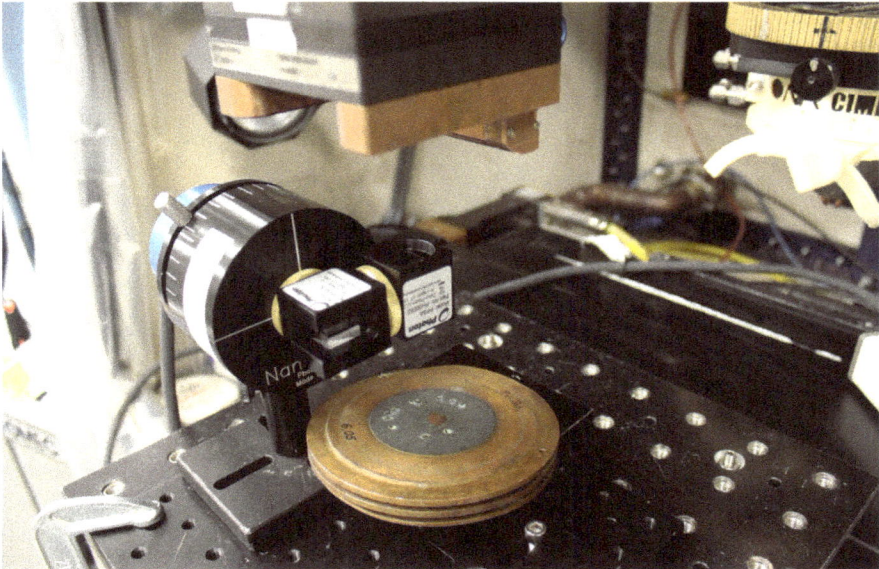

Figure 3. Photograph of the setup used to characterize the focused laser beam.

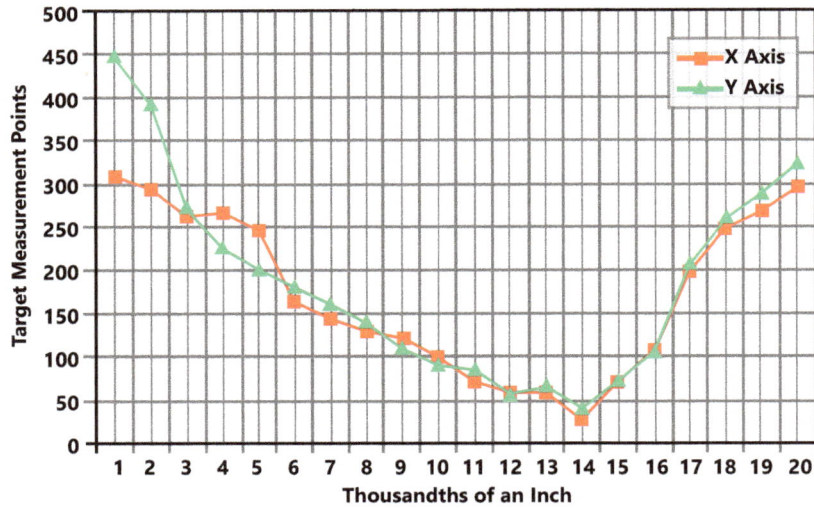

ning slit of the detector for measurement.

The detector first must be set up to handle the power density of a 50-µm laser beam delivering 500 W of power; the calculated density is 25.46 MW/cm². To allow such a high power density onto the slit without causing damage to the measurement device, a dual front-surface beamsplitter was attached to the face of the detector so that, combined with optical wedges, it offers a maximum measurable power density of 100 kW/cm².

Figure 3 shows the experimental setup used to make these measurements. The NanoScan has a dual prism just beneath the optical head of the laser. Beneath the first prism is a 5/8-in.-thick copper plate that was used as a beam dump to absorb the 460 W of laser beam power from the first prism.

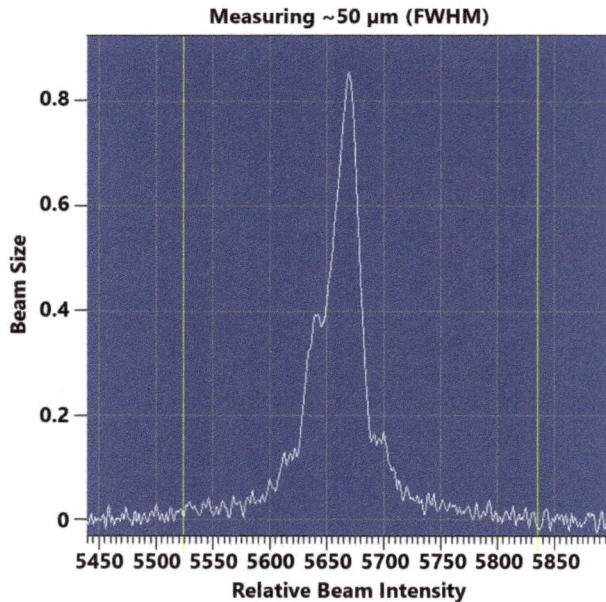

Figure 5. Beam profile near the focal point (in the X direction). FWHM = full width half maximum.

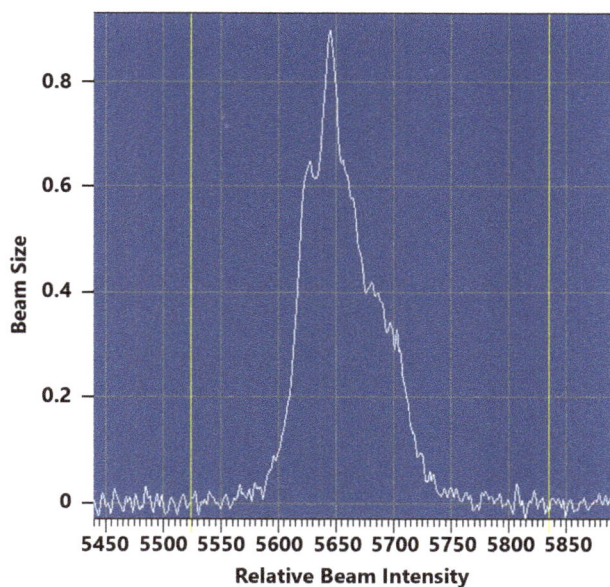

In this setup, the first wedge split was 90 percent or 460 W of the maximum delivered power. The second wedge split was 90 percent of the remainder, or 36 W, where a simple 40-W power meter beam dump was used (not shown).

Figure 4 shows the size of the beam in both the X and Y dimensions taken from each of the prescribed increments. The initial findings showed that the beam was slightly smaller at the focal point than was calculated. In addition, the location of the focal point was slightly below the estimated focal point. It was also determined that the "calculated" location of the focal spot was not exactly where it was expected to be — another key element in the performance of the laser contributing to a quality product. Further testing is needed to verify this information, but the results point to the need for further diagnosis of the laser settings and operation before preparation of production parts.

The measurement procedure consisted of taking 25 separate readings, starting at 0.375 in. above the estimated focal point (identified as zero-zero) and proceeding at increments of 0.03126 in. down to 0.375 in. below the estimated focal point. The chart in Figure 4 shows the plot of the beam size, and Figures 5 and 6 show cross-section measurements of laser beam power near the focal point.

These tests showed that, with this setup, the laser beam power lost between the set point and the substrate was quantified, the laser beam spatial energy distribution was quantified, and the predetermined location of the focal spot was quantified.

With this information, it is possible to define an AM procedure that specifies the power level, spatial energy distribution of the laser beam delivered to the substrate, and the focal spot, ensuring repeatable results.

Meet the author
Dick Rieley is the sales manager for the mid-Atlantic region at Ophir-Spiricon LLC.

Advances Add Up for 3D Printing

Advances in processing, lower costs and improved performance could soon move 3D printing from prototyping to commercial production of aerospace and medical components.

BY HANK HOGAN, CONTRIBUTING EDITOR

For additive manufacturing, aka 3D printing, 2016 could be a very good year. Some big names are entering the field, advances promise better processing and signs point to movement into the mainstream. At the same time, the industry is working on developing standards while continuing to lower cost and improve performance — thereby addressing issues that have hindered adoption of additive manufacturing.

One result looks to be explosive growth. Titanium powder is a raw 3D printing material, and its consumption is a proxy for overall additive manufacturing activity. That appears strong, based on projections for the titanium powder market over the next decade.

"That is about a 31.4 percent compound annual growth rate," said Scott Dunham, senior analyst at SmarTech Markets, a 3D printing analysis firm.

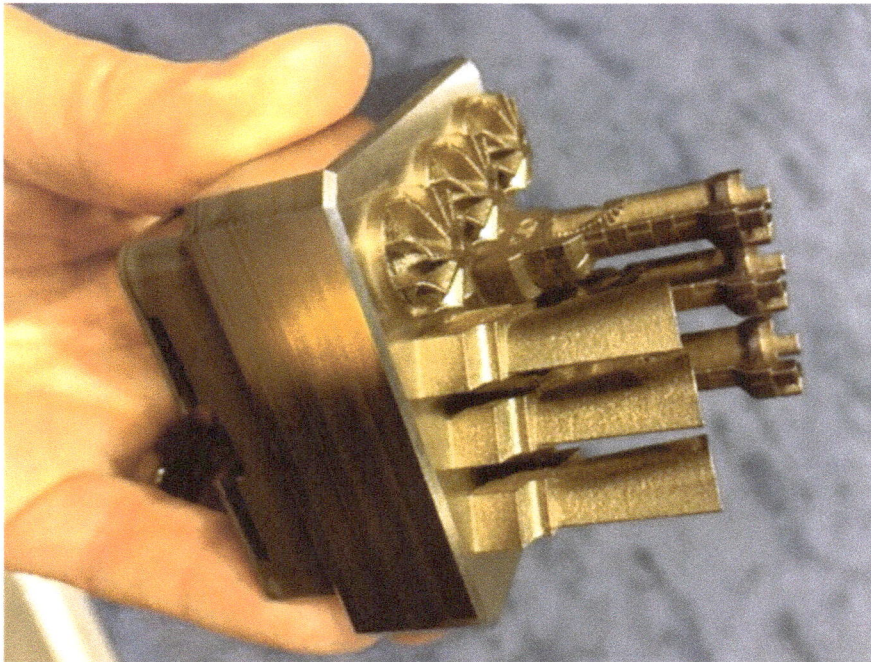

Nine parts additively manufactured on a single sub plate, ready for final subtractive manufacturing via traditional milling, grinding and so on. Courtesy Georg Fisher Machine Systems.

A variety of shapes can be produced by additive manufacturing, and researchers are investigating how to improve this 3D printing process. Courtesy of Siemens.

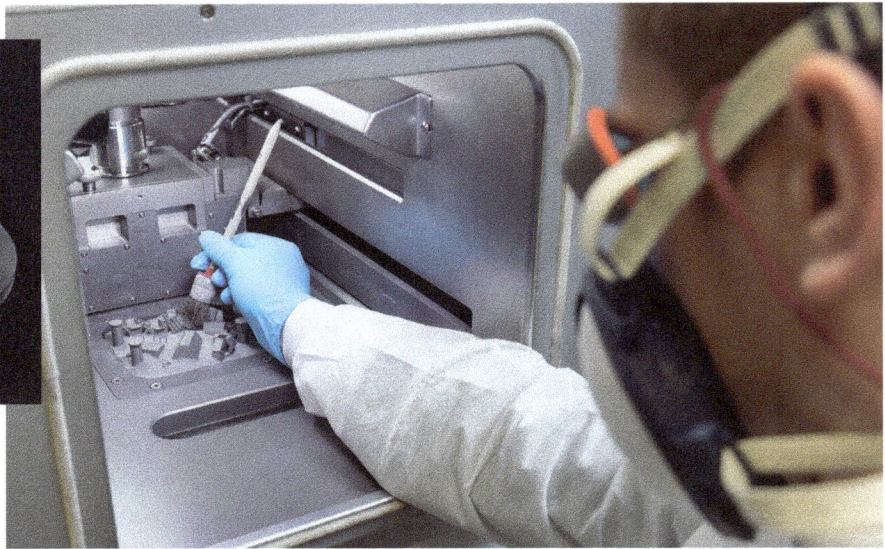

SmarTech Markets forecasts the titanium powder market will grow from $66 million worldwide in 2015 to $776 million by 2024. Titanium is favored in aerospace and medical applications, two areas willing to pay a premium to produce high-performing critical parts. This may be for orthopedic implants, aircraft support structures and engine components. GE, for instance, is making all-important fuel nozzles for its latest engines via additive manufacturing.

What also sets medical and aerospace apart is that both are no longer interested in prototyping alone. "In medical, they're printing a lot of parts for actual final use in patients. In aerospace, there's an impressive number of parts that are scheduled for actual production in the coming months," Dunham said.

An indication of this increasing maturity can be found in what Siemens AG is doing. A diversified technology and manufacturing company, Munich-based Siemens makes, among other things, gas turbines. The company today uses additive manufacturing to create better and higher-performing burner tips for those turbines.

Eventually, a company goal is to have turbine blades produced through 3D printing. That would enable the creation of cooling channels inside the blades, boost-

A child's tracheal splint created by additive manufacturing is shown at right. Laser-sintered parts are customized to the individual. Courtesy of Leisa Thompson, Photograpy/UMHS.

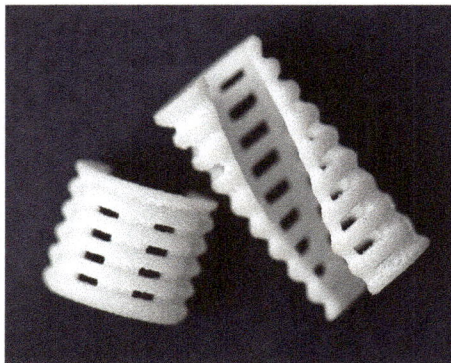

ing their efficiency. However, it's not possible to do so with currently available additive manufacturing due to the stress the blades experience as they spin inside a turbine. Siemens is, therefore, actively engaged in 3D printing-related research and development.

However, the company is not involved in the production of 3D printers, according to spokesperson Peter

Jefimiec. Instead it is striving to be a supplier and technology provider to builders of additive manufacturing machines, thereby helping move the technology into wider use.

"Our aim is to anchor additive manufacturing firmly in the industrial production process, which means industrializing 3D printing," Jefimiec said.

For another data point about the move from niche status to full manufacturing, consider the entry of HP Inc. into the 3D printing market. The Palo Alto, Calif.-based company claims its Multi Jet Fusion technology, which is not photonics-based, offers up to a tenfold increase in speed and better quality than the competition.

Those statements may be widely and independently confirmed once the printers hit the market some time this year. Even before that, though, the fact that a company with a long history as a dominant player in the printer market will now be selling additive manufacturing products is significant in terms of validating the whole field.

Count Andrew Snow among those who believe this to be the case. Snow is senior vice president of Novi, Mich.-based EOS of North America, a subsidiary of Germany's EOS. The company makes laser-based additive manufacturing systems and saw a more than 50 percent jump in revenue from 2014 to 2015, growth that Snow attributed to the widening adoption of industrial 3D printing in aerospace, medicine and elsewhere. HP's entry could well help continue that usage climb.

Another reason could be ongoing additive manufacturing improvements. EOS, for example, is deploying real-time monitoring, using photodiodes to look at the melt pool. This will be done along the axis of the laser and also from an off-axis vantage. Such monitoring can provide a wealth of information.

"Not only does it examine the mechanical quality of the components being produced but also you're continually monitoring the overall robustness and reliability of the hardware as well," Snow said.

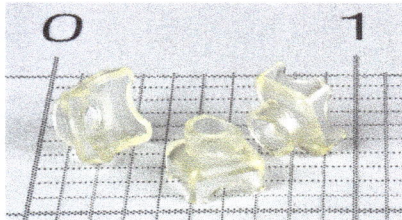

A two-photon polymerization technique enables 3D printing of a miniature replica of the Roman Colosseum (left) as well as lens holders (top). Courtesy of Nanoscribe.

A laser control scheme enables tailoring of the microstructure of materials of additive manufactured parts, such as these blades from Taiwan's Industrial Technology Research Institute. Courtesy of ITRI.

At the moment, the technology is only being used to collect information on the various material and processing parameters. Such in-situ monitoring could someday be used for part certification or in a process control loop.

Another reason for future growth may be continued development of industry standards governing 3D printing technology, design and materials. EOS, for instance, is active in the relevant committees in ASTM International, the global standards-setting body. Among other things, standards can help in designing for manufacturability. Today, this information is making its way into academic settings and the training of the next generation of engineers and designers, according to Snow.

In speaking of the future, he noted that no single material is suitable for all applications. Consequently, it will likely be true that both additive and subtractive manufacturing will be used in industrial processes in the future. As a result, there needs to be ways to integrate a variety of techniques, enabling the passing off of parts from additive to subtractive machines or vice versa with little to no human intervention. EOS is working on developing such methods, Snow said.

Nanoscribe GmbH, in Eggenstein-Leopoldshafen near Karlsruhe, Germany, takes a different approach than other companies in its additive manufacturing systems, with the result being that it can create objects with a feature size as small as 100 nanometers, said CEO Martin Hermatschweiler. This is possible because the technique uses a two-photon process to harden a photoresist. As the name implies, this only happens when two photons are simultaneously absorbed within a small volume of photosensitive polymer.

"It's like having a very precise pen in your hand, and with this pen you can define where you want to polymerize the resist and do it nowhere else," Hermatschweiler said.

The resulting structures are smooth enough to be used as an optical surface without any polishing or other post-processing. The technique is largely limited to polymers or hydrogels, although these can be functionalized by, for example, adding nanoparticles that react to the presence of chemicals or biological compounds.

Once made, the parts might be used as a mold or stamp to create others, which cuts costs. Another way to reduce manufacturing expenses would be to polymerize the skin of the resist and then use flood illumination to finish the interior.

These are examples of the ongoing trend of falling costs in additive manufacturing, one reason for its greater acceptance. Another is that systems are increasingly easy to use. This can be seen in Nanoscribe's products by the fact that you don't need to be an expert to run them, according to Hermatschweiler.

He thinks that the entry of large well-known industrial companies will further boost the additive manufacturing field. One way could be the development of new standards, such as is being done by the Wakefield, Mass.-based 3MF Consortium. Founded in 2015, the organization counts HP, Microsoft, Siemens and others among its members. Its goal is to define a 3D printing format to allow designs to be easily sent to applications, platforms, services and printers.

A final instance of a technological advance comes from the Hsinchu, Taiwan-based Industrial Technology Research Institute. Researchers there have developed the ability to use lasers to control the microstructure of the material in additive manufacturing. This is done through a complex beam shape with many irradiation modes that controls thermal variations in the 3D printing process, according to Ji-Bin Horng, senior principal engineer.

This control can create unique and useful properties in materials because it allows the adjusting of parameters on a minute scale, Horng said. "A turbo blade with directional microstructure will have durability 100 times higher than normal equiaxial grains. Material microstructure of key components has an optimal choice for an application."

The approach, which is based upon what is called an optical engine, has been demonstrated in metal structures but it can be applied to ceramic, polymer and composite materials, Horng said. An advanced additive manufacturing technology, it is only in the early stages of commercialization. There has been a demonstration of the concept but no actual products produced. The optical engine could potentially be installed as an add-on to existing 3D printers, provided that they are laser-based.

One benefit of the approach is that it could help protect intellectual property. Instead of everything about a part that makes it useful being embedded only within its size and shape, some of that might be found within the material itself. For instance, the hundredfold increase in durability arising from aligning metallic grain structure would come from the manufacturing process, not the part geometry.

The extra degree of freedom derived from being able to adjust material properties could be the next advance in additive manufacturing. It would be part of a change in how designers think about parts and products.

As Horng said, "Engineers should consider the mechanical properties everywhere in the part, simulate the best design and set the related parameters."

> The extra degree of freedom derived from being able to adjust material properties could be the next advance in additive manufacturing.

Laser Microfabrication Techniques Move Rapid Prototyping to the Mainstream

With some methods already accepted as mature process technologies, laser microfabrication continues to replace complex fabrication protocols.

BY BEDA ESPINOZA, TAGUHI DALLAKYAN AND PHONG DINH,
MKS INSTRUMENTS LIGHT & MOTION DIVISION

Rapid prototyping is a fabrication technique that uses additive layer-by-layer fabrication to create three-dimensional structures from devices ranging in scale from MEMS devices to 3D-printed houses. The technique has its origins in the mathematical theories of Herb Voelker[1] that provide the basis for 3D computer-aided design (CAD)[2]. Carl Deckard applied these principles in the 1990s to develop a fabrication method that he called "selective laser sintering[3]," which rapidly evolved into 3D printing, rapid prototyping and other laser microfabrication methods[4].

Recently, rapid prototyping has captured the interest and imagination of the

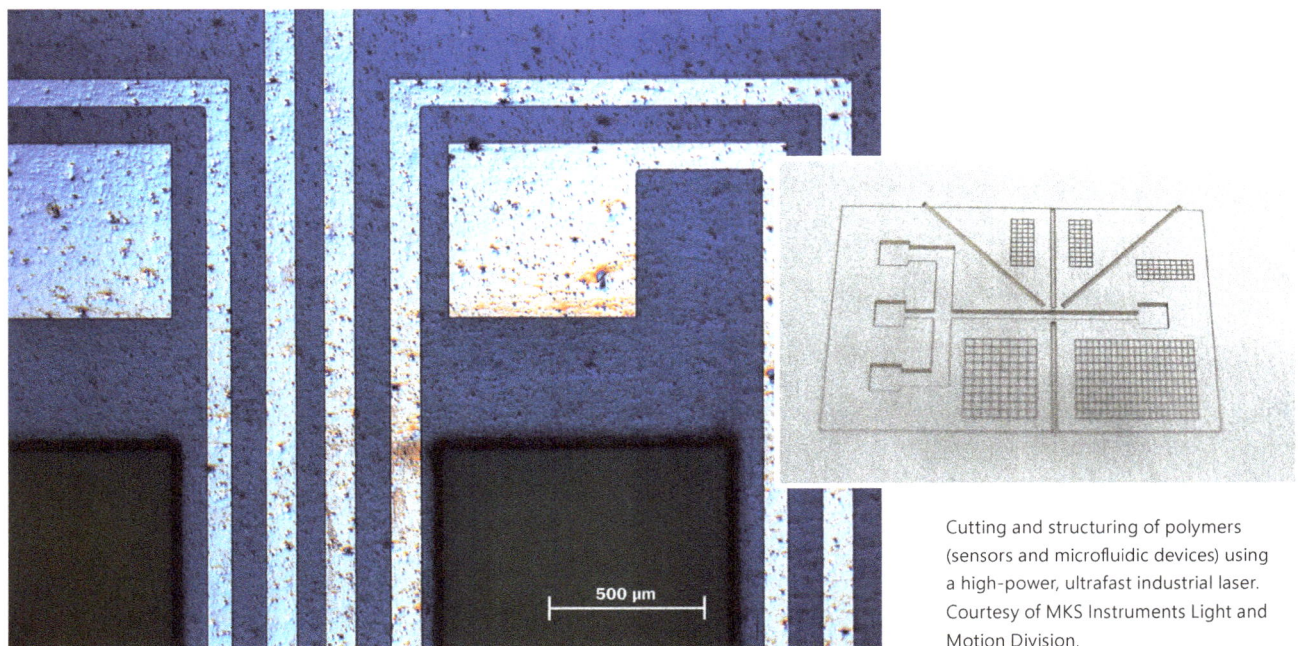

Cutting and structuring of polymers (sensors and microfluidic devices) using a high-power, ultrafast industrial laser. Courtesy of MKS Instruments Light and Motion Division.

public owing to the potential for easier and cheaper access to everything from replacement parts for common consumer products to dental appliances[5]. President Barack Obama, in his 2013 State of the Union address, promoted rapid prototyping as having the "potential to revolutionize the way we make almost everything." Rapid prototyping machines are now common in many settings, from private homes to the engineering departments of major corporations.

According to D.V. Mahindru and Priyanka Mahendru[6], there are at least six different rapid prototyping techniques commercially available:

- stereo lithography,
- laminated object manufacturing,
- selective laser sintering,
- laser beam machining,
- solid ground curing, and
- 3D inkjet printing.

All of these techniques have certain basic steps in common. A 3D CAD model must be developed and converted to a stereo lithography format model that can be sliced into thin cross-sectional layers. The geometrical information from these layers is used to control equipment that performs the operation — laser ablation, powder fusion, polymerization, additive printing — necessary to construct the artifact on a one-layer-at-a-time basis.

Rapid prototyping tools for micromachining and microfabrication benefit from high-intensity, ultrafast femtosecond lasers that produce extremely localized heating and no collateral damage or heat effects in surrounding materials — a minimal heat affected zone (HAZ). Along with an ability to focus the beam to a small (1 μm) spot size, the minimization of adjacent heat-affected zones is fundamental for the production of high-quality three-dimensional micro- and nanostructures. Optical components are needed that manage the laser beam from the source to the material being worked as is software to translate 3D CAD formats into executable machine instructions. Ancillary components including power meters, beam

Surface structuring of AlO_2 ceramics using a femtosecond laser. Right image with gold coating on the surface. Courtesy of MKS Instruments Light and Motion Division.

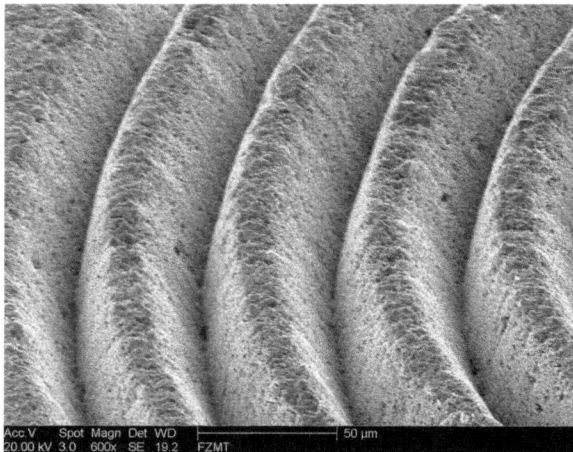

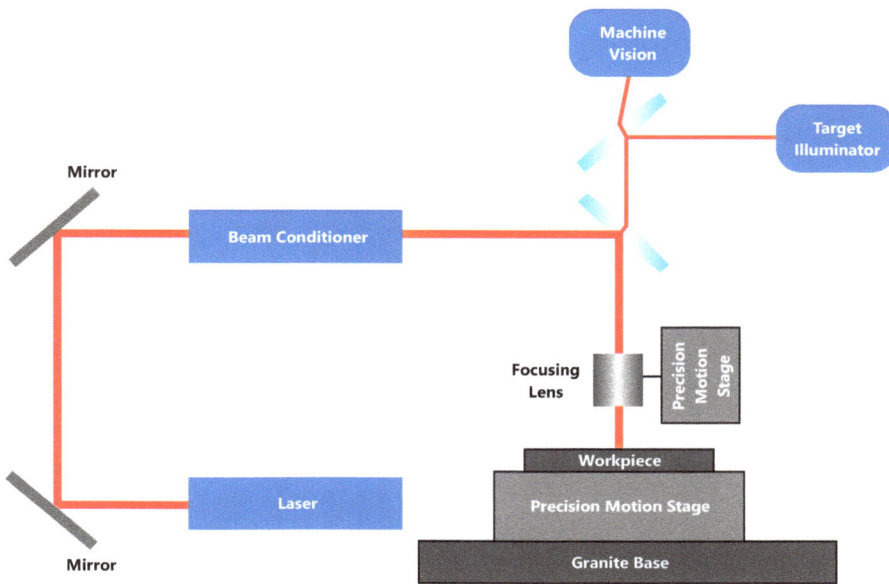

Figure 1. Components in a laser micromachining tool. Courtesy of MKS Light and Motion Division.

profilers, galvo scanners (control mirrors), etc., ensure repeatability in the optical characteristics governing the process. Motion stages offer fast, accurate and precise positioning of both the laser beam and the substrate at the nanometer scale, determining the precision with which microfeatures can be added or removed in the fabrication process and the manufacturing time requirements. Finally, controllers are required to drive the motion stages and synchronize their motion with the laser pulses and the ancillary devices.

Notably, rapid prototyping systems can be configured using "off-the-shelf" lasers, optical software and motion control tools.

Laser micromachining

Laser micromachining is a subtractive machining process in which the thermal energy of a laser is used to ablate or otherwise alter metal or nonmetal surfaces. The basic components of a laser micromachining tool include a device for scribing and ablation processes on a 6-in. silicon wafer (Figure 1). The tool contains optical and motion control subsystems that are synchronized and controlled using a computer running specialized laser micromachining software.

A wide variety of lasers are employed in advanced micromachining applications. One example is an automated, high-power (10- to 20-W) fast-pulsed laser such as the Spectra-Physics Spirit industrial femtosecond laser. Adjustable pulse energy and repetition rates (up to 40 μJ and 1 MHz, respectively) allow precise control of the energy delivered to the workpiece, minimizing the HAZ. The optics system consists of the laser, galvo scanners, a beam collimator and a focusing lens, along with the lenses and illumination needed for observing the workpiece.

The motion control system is mounted on a granite base to ensure stability and repeatability in the movement and positioning of the workpiece. The system uses

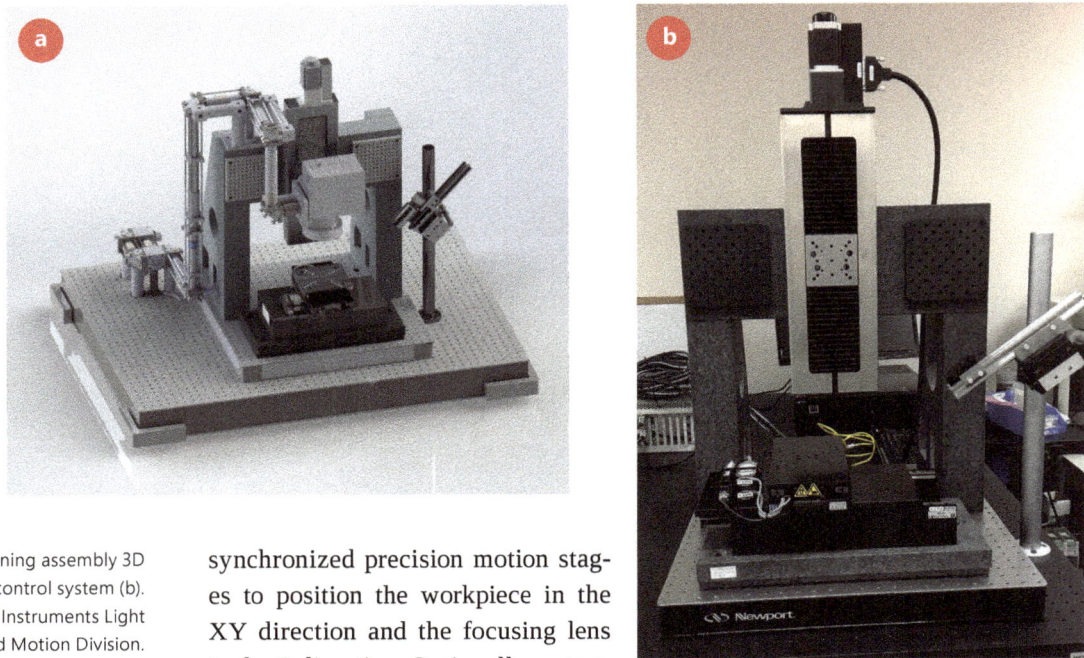

Figure 2. Laser machining assembly 3D schematic (a); motion control system (b). Courtesy of MKS Instruments Light and Motion Division.

synchronized precision motion stages to position the workpiece in the XY direction and the focusing lens in the Z direction. Optionally, a stage for rotary positioning of the workpiece can be stacked on the XY stage. For micromachining applications, these stages require positioning accuracy on the order of ±0.5 µm with a repeatability of <0.05 µm in XY directions. Positioning accuracy requirements in the Z direction are somewhat more forgiving at <5 µm, with resolution specifications typically about 0.1 µm. Travel range in the XY directions for applications such as the one in this discussion (the workpiece is a 6-in. diameter silicon wafer) is minimally 160 mm (slightly greater than the workpiece diameter) with minimal incremental motion of 0.010 µm. Translational speeds on the order of 300 mm/sec and accelerations of up to 5 m/sec^2 are required in each axis to ensure reasonable throughput. Motion of the workpiece and lens is controlled and synchronized using an integrated motion controller/driver to control movement in the X, Y and Z directions (and rotational motion, where equipped). The system requires specialized software for laser machining and machine vision. This software translates the 3D CAD model into adjustments to the stages, galvo scanners, optical components laser sources, sensors, power meters, etc.

Figure 2 shows a 3D drawing of a laser micromachining system, along with an image of a real-world motion control system for use in such a tool. The design employs a dual optical system to allow special beam routing for two wavelengths and beam collimation depicted in the 3D schematic (Figure 2a). These types of systems have been used to produce microstructures such as optical waveguides.

Ablative laser micromachining is considered mature technology in a number of industries. Automotive applications include the fabrication of fuel injector nozzles, part marking and microwelding. In consumer electronics, laser micromachining is employed in applications such as glass cutting and structuring for OLEDs.

Additive micromanufacturing

Three-dimensional additive microfabrication is the opposite of laser micromachining in that it creates a microstructure using layer-by-layer material addition through sintering, or some other mechanism of solidification of a source material, rather than by removing material from an already solid substrate. In additive microfabrication, the laser beam provides the energy necessary for patterned, layer-by-layer fusion of solid materials (sintering) or for a patterned, layer-by-layer polymerization (stereolithography or two-photon polymerization)[7,8]. As with laser micromachining, additive microfabrication can benefit from femtosecond lasers to minimize HAZ and, in the case of two-photon polymerization, to deliver the high local photon densities necessary to induce localized polymerization.

Two-and-a-half and three-dimensional additive micromanufacturing is finding applications in the fabrication of complex microstructures in many areas, including electronics, microfluidics, MEMS, micro-optical electro-mechanical systems (MOEMS), biomedical, chemical analysis and decorative surface structures, to name a few. It can produce these microstructures from a wide variety of materials, ranging from metals and alloys to ceramics and polymers. For many applications, additive micromanufacturing is superior to conventional photolithographic patterning for micro- and nanoscale device fabrication since it can create high-aspect-ratio structures with fully vertical walls, re-entrant angles and other free-form structural characteristics that are impossible with conventional patterning technology. Additionally, additive microfabrication offers a more direct simplified route to such structures.

Configuring an additive micromanufacturing tool uses similar components to those employed in laser micromachining. A wide range of lasers can be used, including UV pulsed nanosecond lasers and industrial femtosecond lasers such as the 1040- and 520-nm Spirit lasers. XY and Z motion stages coupled with a granite base and bridge provide the necessary level of precision and accuracy in motion control. Additive microfabrication systems can be controlled using LMS Lite software equipped with a 3D module. Ultrafast manufacturing can be achieved through the use of controllers and drivers for high-speed multi-axis trajectory pulsing.

Additive microfabrication is finding practical application in areas such as MEMS for the automotive industry; microfluidics for chemical, biochemical and pharmaceutical industries; and component fabrication for optoelectronics.

A Laser μFab

The commonality of equipment requirements for ablative laser micromachining and additive microfabrication shows that the integration of these two micromanufacturing techniques into a single manufacturing tool is feasible. The commercially available μFab is one tabletop tool that combines these capabilities (Figure 3). This tool, configured with a 1040-/520-nm industrial femtosecond laser, with either XMS or VP series motion stages and an XPS-Q4 four-axis controller, can

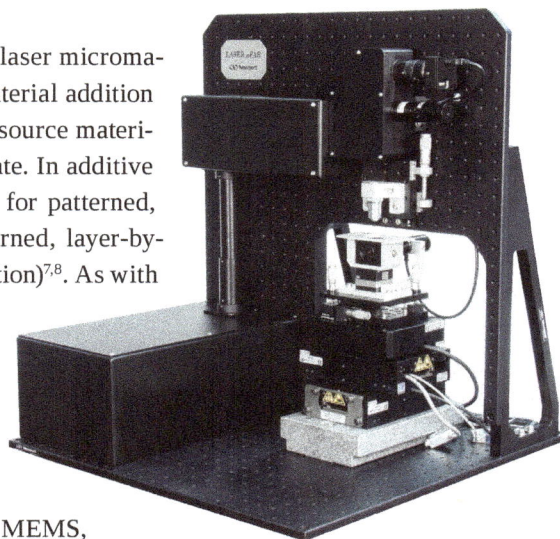

Figure 3. An integrated Laser μFab from Newport. Courtesy of MKS Instruments Light and Motion Division.

a

b

c

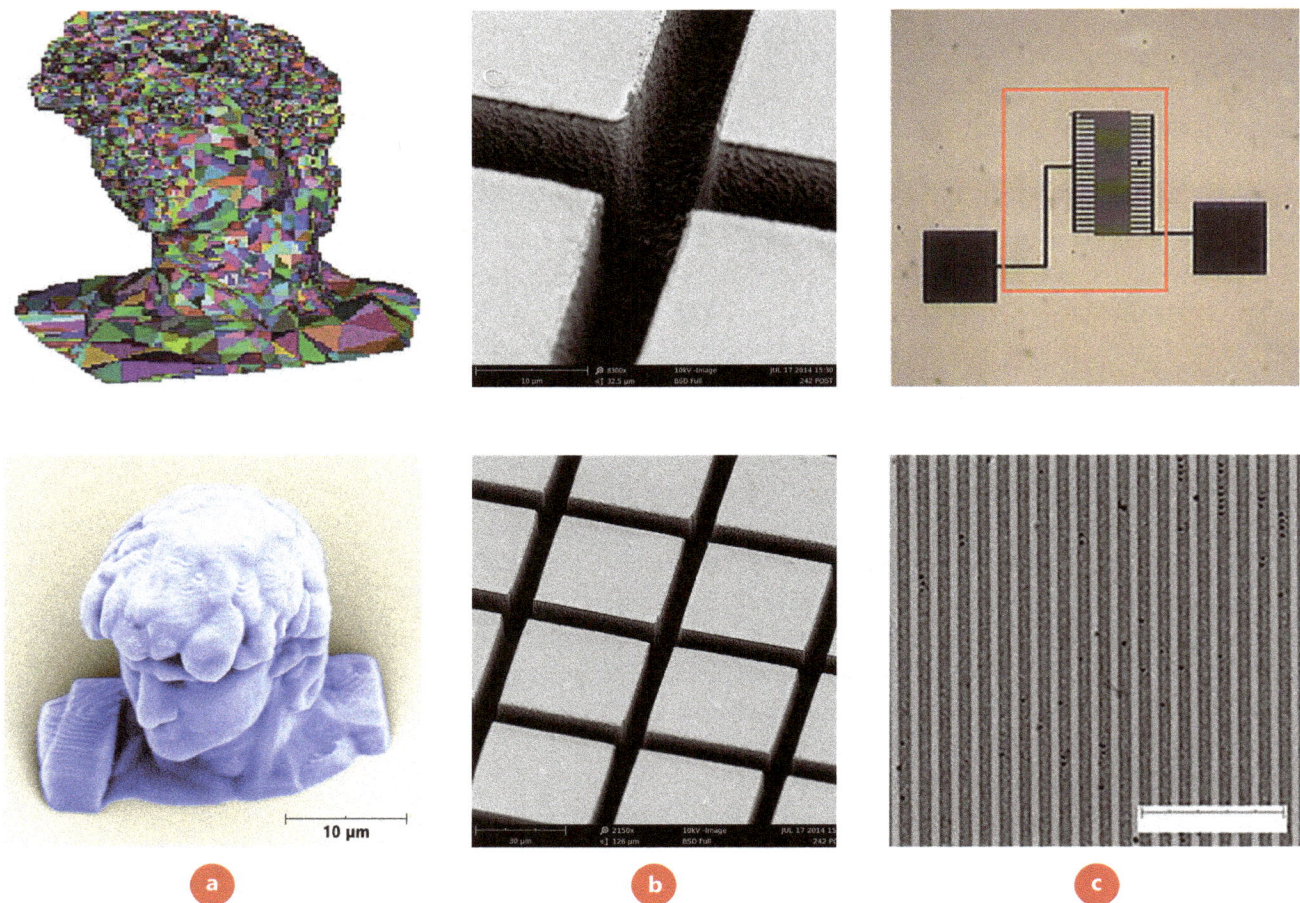

Figure 4. Microfabrication using the Laser µFab: STL file and 3D microstructure produced using TPP (10-µm scale) (a); 2D patterns on a polymer produced using laser ablation (5-µm channels) (b); laser-assisted silver deposition on glass (3.5-µm line widths) (c). Courtesy of MKS Light and Motion Division.

process glass, metals, polymers, semiconductors or ceramics using transformative or ablative microfabrication, in addition to additive processing techniques such as two-photon polymerization and laser-assisted deposition. On-axis reflective light microscopy is used for monitoring the laser processing in real time. Customizable software controls system operation.

The tool can be utilized as an additive microfabrication tool (TPP), a laser micromachining tool and a tool for laser-assisted deposition to fabricate microstructures (Figure 4). Different laser microfabrication applications are appropriate for certain components.

Laser microfabrication has evolved rapidly since the 1990s, to the point where many methods have become accepted as mature process technologies in a variety of industries. These methodologies continue to find new applications in replacing more complex fabrication protocols or in entirely new technologies.

Meet the authors

Phong Dinh is an applications engineer, Taguhi Dallakyan is a product specialist and Beda Espinoza is a product marketing manager for the MKS Light and Motion Division, located in Irvine, Calif.

References

1. H. Voelcker, et al. (August 1978). The PADL 1.0/2 system for defining and displaying solid objects. *Comput Graph (ACM)*, pp. 257-263.

2. C.M. Eastman and K. Preiss (1984). A review of solid shape modelling based on integrity verification. *Comput Aided Des*, Vol. 16, Issue 2, pp. 66-80.

3. J.J. Beaman and C.R. Deckard (July 3, 1990). Selective laser sintering with assisted powder handling. U.S.A. Patent US4938816 A.

4. X. Yan and P. Gu (1996). A review of rapid prototyping technologies and systems. *Comput Aided Des*, Vol. 28, Issue 4, pp. 307-318.

5. H. Lipson and M. Kerman (2013). *Fabricated: The New World of 3D Printing*. Indianapolis: John Wiley & Son.

6. D.V. Mahindru and P. Mahendru (2013). Review of rapid prototyping technology for the future. *GJCST, Graphics and Vision*, Vol. 13, Issue 4.

7. Application Note: Three-dimensional microfabrication by two-photon polymerization [online]. Available: https://www.newport.com/medias/sys_master/images/images/h26/hf7/8797287743518/Three-Dimensional-Microfabrication-App-Note-37.pdf.

8. Application Note: Workstation for laser direct-write processing [online]. Available: https://www.newport.com/medias/sys_master/images/images/h15/h4a/8797304651806/Workstation-for-Laser-Direct-Write-Processing-App-Note-39.pdf.

Two-Photon Polymerization: Additive Manufacturing From the Inside Out

Fabricating microstructures with submicron feature sizes using a 3D writing technique.

BY TOMMASO BALDACCHINI, MATTHEW PRICE AND PHONG DINH, NEWPORT CORP.

The 20th century was an era of revolutionary materials science development, as plastics, semiconductors and biomaterials were introduced, and established materials such as metals and ceramics were significantly upgraded. These new and improved materials ushered in countless applications in industries as diverse as electronics, transportation, energy and health care.

In addition to discovering new applications for materials, existing applications can also be improved considerably through better materials. In particular, the desire for lighter or smaller materials — without a decrease in functionality — is a challenge for today's scientists and engineers. One need only look at the computer industry to see how materials technology has contributed to reducing the size of a machine that once filled an entire room to fit into the palm of one's hand.

The impending breakthrough applications of the 21st century demand much more than just lighter or smaller materials. They must be stronger, stiffer, capable

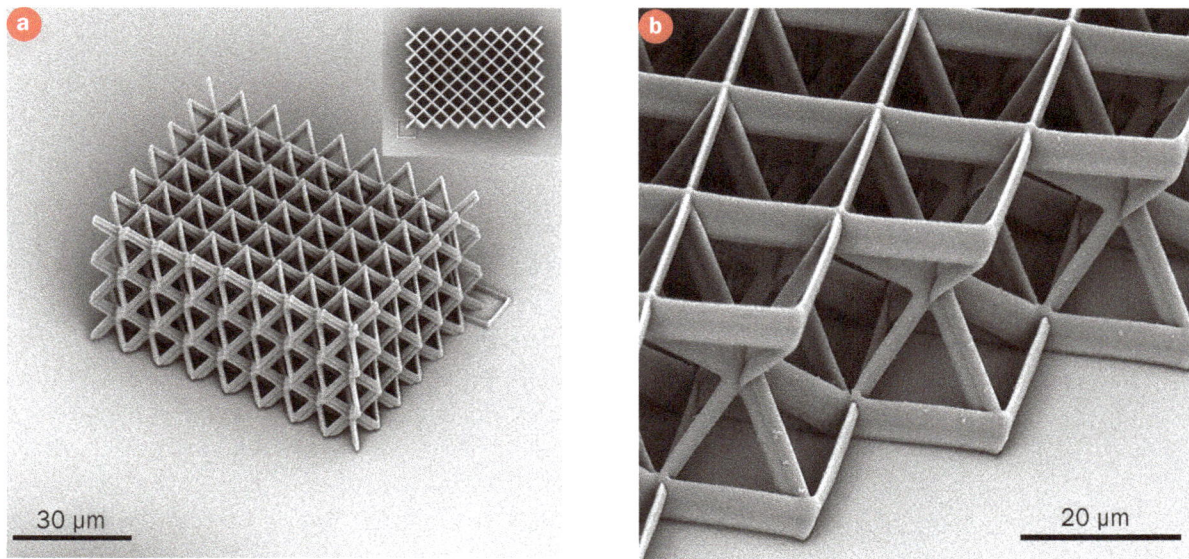

Figure 1. Scanning electron microscope (SEM) images of microstructured materials fabricated by two-photon polymerization (TPP). In (a), the whole pattern is shown at an angle and from a top view (inset). In (b), a magnified view of the unit cell of a microstructured material is shown. Images courtesy of Newport Corp.

30 μm

20 μm

of damping a range of vibrations or managing heat. Traditional materials science, which has focused mainly at the chemical level, is a mature field and therefore not likely to provide the solutions needed for new applications. Thankfully, a new technology is emerging: microstructured materials.

The concept of a microstructured material can be seen in the macro world. Lattice towers and truss bridges, for example, are dependent not only on the type of materials chosen, but also on the size, shape and placement of the components that make up the entire structure. With a clever design, a tower or bridge can be built with the minimum amount of materials to withstand heavy forces and loads, making them lightweight and strong at the same time.

At a microscopic, or cellular, level, a cross section of a microstructured material also looks like an array of trusses (Figure 1). As with a macrostructure like a lattice tower, these trusses can also be designed to provide the material with additional features. Moreover, the architecture of microstructured materials can advance the properties of the ultimate object far beyond what traditional materials science has done at the chemical level. Proper design of microstructured materials can provide more than one functionality, such as high stiffness and damping coefficient. Microstructure design can also decouple properties that typically compete with each other, so it is possible, for example, to have a single material that provides high strength at low density.

Microstructured materials are an exciting field, and, like most potentially disruptive technologies in their early stages, they also present challenges. In particular, the manufacture of these complex geometries at the microscopic level is one of the first hurdles researchers must overcome.

Two-photon polymerization manufacturing

A promising manufacturing technique for the realization of microstructured materials is two-photon polymerization (TPP). There are two key reasons that render TPP an attractive solution for the fabrication of microstructures. The first is that TPP is intrinsically a 3D writing technique; it does not require a layer-by-layer approach to create complex objects. TPP can also create microstructures with submicron feature sizes in a straightforward manner. These characteristics originate from the nonlinear optical nature of light absorption in TPP and from the chemistry of the polymerization.

In a typical experiment, near-infrared (NIR) emission is used to excite a photosensitive material (resin) that upon light absorption undergoes a phase change from liquid to solid through a polymerization process. Since the resin is transparent in the NIR region of the spectrum, high numerical aperture (NA) lenses and femtosecond pulsed lasers are employed to increase the probability of a multiphoton absorption event to occur. That is, two or more photons are simultaneously absorbed by specialized molecules in the resin (photoinitiators) to create the active species that start polymerization. Under these conditions, multiphoton absorption occurs only in the region where light intensity is the highest. That confines polymerization within the volume of the focused laser beam (voxel). Three-dimensional

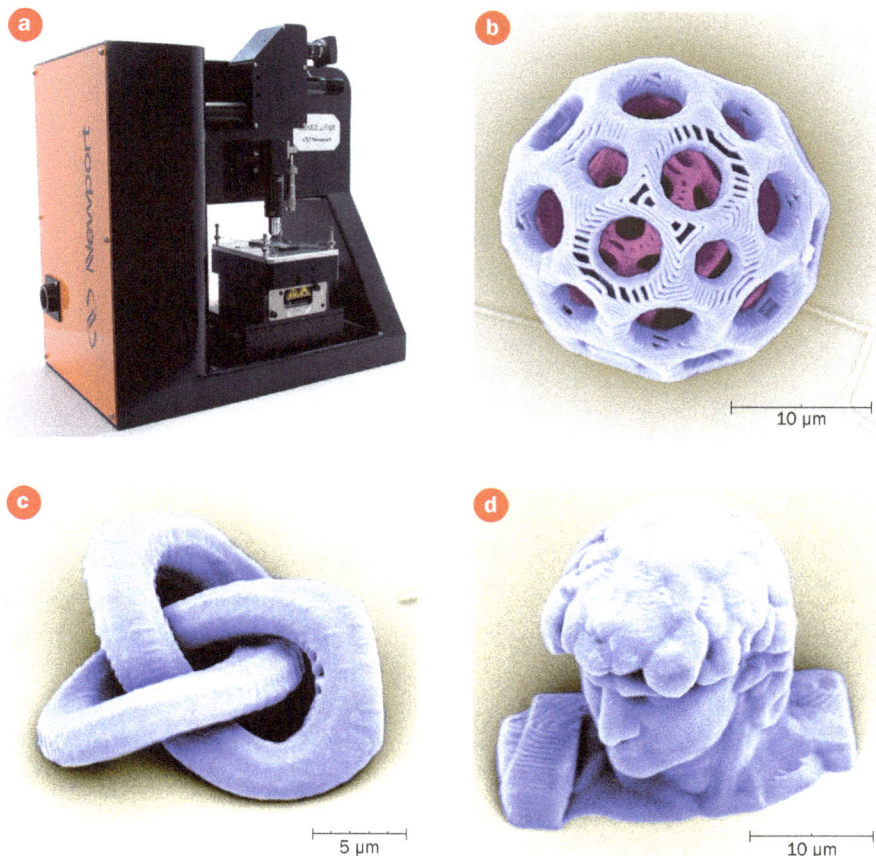

Figure 2. (a) Newport's Laser µFAB. (b,c,d) False color SEM images of three-dimensional microstructures fabricated by TPP using the Laser µFAB.

microstructures are created by precisely overlapping voxels through the scanning of either the laser beam or the sample around predetermined geometries. Successively, the microstructures are revealed by washing away the unsolidified part of the resin using an organic solvent.

TPP offers a unique combination of advantages:

- No topological constraints are present in the fabrication of 3D structures.
- Subdiffraction-limited feature sizes can be attained by employing laser intensities just above the intensity threshold at which polymerization will occur.
- Movable components can easily be fabricated without the use of sacrificial layers.
- The carbon-based nature of the photosensitive materials can be used as a chemical handle to fabricate structures with tunable physical and chemical properties such as hardness, shrinkage, index of refraction and chemical specificity.

It should not come as a surprise then that 3D microfabrication by TPP has evolved in recent years from a novel technique employed mostly by laser specialists to a useful tool in the hands of scientists and engineers working in a wide range of research fields. A perusal of the scientific literature on this topic reveals how extensive the impact of TPP has been in applications that require high-precision

3D writing — from the fabrication of optical and mechanical metamaterials to the development of rationally designed substrates for cell motility studies.

At Newport's Technology and Applications Center, scientists have developed the Laser μFAB that can be used to fabricated 3D microstructures of unprecedented complexity and with great finesse (Figure 2). A set of high-precision stages are used within the Laser μFAB to move the sample across a fixed laser excitation beam. Although this method of performing TPP allows for patterning large areas that are limited only by the total travel distance of the stages, it also requires the highest performances by the stages. Consequently, the development of such a system involves unique challenges, but solutions are available.

Motion control

The spatial tolerances required by TPP fabrication can be found directly in scanning electron microscope (SEM) images with feature sizes at the submicron scale and tolerances measured in nanometers. While SEM images provide a clear picture of the spatial tolerances, the dynamic tolerances are not always readily transparent. These dynamic demands come in part from the nonlinear nature of the TPP process and are especially evident in the uniformity of periodicstructures in 3D lattice arrays and feature sizes near the TPP process limit.

Fabrication of three-dimensional periodic microstructures is especially demanding because the motion system must demonstrate repeatability in six degrees of freedom to the nanometer scale. This puts specific demands on flatness and straightness as well as Abbe error induced from parasitic angular motion, such as pitch and yaw deviations. TPP processes using a high NA microscope objective focused spot are also characteristically slow. This means that the impact of motor heating over extended operating times should be negligible. This also introduces constraints related to in-material focus and the flatness of the stage to hold process planarity and correct optical axis height.

The slow process speeds of TPP, combined with the nonlinear nature of the TPP process itself, contribute an extra set of very strict dynamic demands. Precise

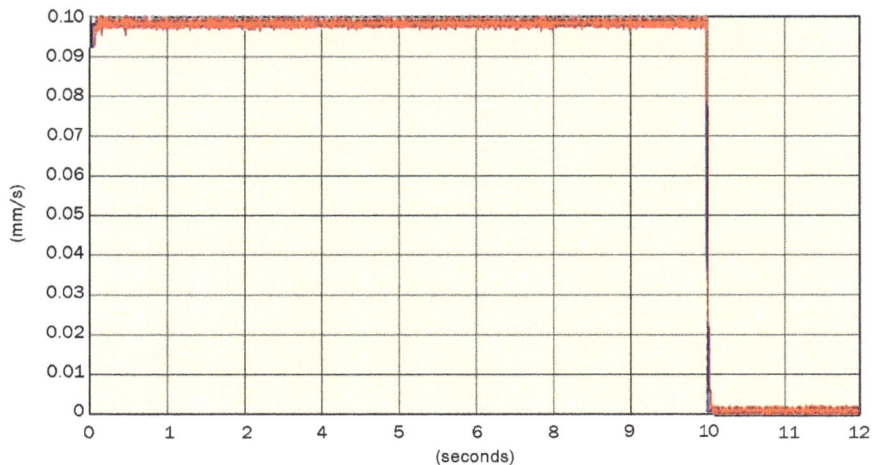

Figure 3. Dynamic ramp and velocity characteristics of scanning XM stage tuned for the TPP process.

dosing requires a uniform process velocity to achieve uniform microstructures. This includes coordinated motion between stages while operating on multiaxis-trajectories.

The in-house materials process and motion experts at Newport work together in the lab to optimize the motion system for the TPP process. The unique combination of multi-mode expertise results in motion systems with optimal process-specific PID (proportional, integral, derivative) tuning, inertia compensation and dynamic motion profiling that compensates for second- and third-order dynamic effects as observed in the process results. This combined expertise is delivered in every motion system involved in laser materials processing.

For TPP, Newport uses XM Series linear motor stages. Machined from an airframe aluminum alloy with high tolerance proprietary machining processes, XM stages are both light and stiff. Capable of submicron flatness/straightness over 50 mm of travel and repeatability of 40 nm over those same dimensions, XM stages excel at performance levels demanded by TPP. Combining a frictionless, direct-drive system with a large center-driven thermally decoupled ironless linear motor and high-resolution direct red encoder, XM Stages achieve <0.1 percent velocity stability, negligible heat impact at the material and general performance characteristics to make uniform, repeatable structures to 100-nm feature sizes.

Beyond motion control

One way to further improve the prospects of TPP manufacturing is with more advanced laser-synchronized motion. A challenge for laser processing is synchronizing the firing of a laser with the motion path. Typical lasers are fired based on an internal clock with very short pulses, so attempting to synchronize this with an external motion signal can lead to missed pulses and substandard results. Newport has developed a unique methodology for synchronizing the triggering of the laser with a multidimensional motion path with less than 100 ns of latency, with up to 12 MHz pulse rates, allowing accurate pulse separation/overlap at the submicron scale while moving at speeds on the order of meters-per-second. The result is optimal throughput with highly accurate and repeatable features.

Microstructured materials are a new frontier in technology with the potential to bring innumerable and unimaginable benefits. This will only be achieved through full-scale commercial TPP manufacturing, and one key to realizing this scale is submicron motion control. Fortunately, the precision motion control industry's 50 years of research and development are well-positioned to meet the challenges.

Meet the authors

Tommaso Baldacchini is a staff scientist with Newport Corp., a subsidiary of MKS Instruments Inc., in Irvine, Calif. Matthew Price is a senior business development manager with Newport Corp. Phong Dinh is a senior applications engineer with Newport Corp.

Laser Quality Matters in Additive Manufacturing

The power density profile of the delivered beam must be measured to known standards.

BY KEVIN D. KIRKHAM, OPHIR-SPIRICON LLC

3D computer-aided design (CAD) and additive manufacturing have radically altered how prototype, developmental and customized mechanical components are created. Now the landscape is again changing as direct laser melting, selective laser sintering or metal 3D printing quickly become widely used for critical, customizable or hard-to-fabricate constructs.

Additive manufacturing starts as a high-power fiber laser beam is directed onto a table of metal powder. The laser beam then draws the net shape of the component, melting a few tens of microns thickness at a time as the 3D CAD model is transformed into a durable, accurate and reproducible mechanical component.

Metal powder must be accurately laid across the build area to an exacting thickness while a focused laser beam of known dimension, power and focal spot location is directed to construct the net shape one thin layer at a time. To ensure the metal is completely reflowed — creating the strongest, most homogeneous structure possible without overheating portions of the construct — the power density and location of the focused laser beam must be consistently known.

Data describing how the laser beam focuses to achieve the operational envelope should be analyzed before and after any critical part is made. While laser power and beam profile measurements have become ubiquitous in this age of using lasers for everything from mosquito abatement to personal electronics security systems, measuring the actual beam that interacts with the powder in these systems is anything but straightforward.

To understand the parameters of the process, it is important to know the power density profile of the laser beam. Total power, focused spot size and focal plane location describe the working beam, but these parameters can change as beam delivery optics heat up over time.

Spiricon's HP-FSM-PM combines a 600-W power sensor and CCD camera to measure the focused beam at the work surface. Courtesy of Ophir-Spiricon.

Numerous measurement systems are deployed to try to track these variables and to attempt to constrain the variations so that consistently accurate constructs may be achieved.

Camera, scanning pinhole and noncontact, Rayleigh scatter-based beam profilers measure the delivered beam. These beam profiling systems can help the AM system user map the beam in space, but a National Institute of Standards and Technology (NIST) traceable power/energy measurement device must also be employed so that the power density profile of the delivered beam can be measured to known standards. Hybrid laser measurement systems that include power or energy measurement, beam profile and beam location sensors are needed. Such systems accurately assess the power density profile of the delivered beam as well as the location and consistency of these parameters.

Hybrid systems that include camera-based beam profiling and a laser power sensor help users of additive manufacturing systems understand the process. The camera is precisely located at the build plane so that an accurate power density model of the working laser beam, at the working plane of the additive manufacturing machine, can be made. These hybrid systems measure focal spots from 37 µm to 2 mm and laser power up to 600 W, providing the user with real-time measurements.

Meet the author

Kevin D. Kirkham is the senior manager of product development with Ophir-Spiricon in North Logan, Utah.

Surface Analysis

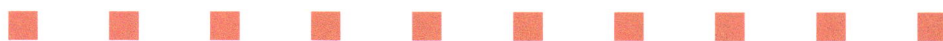

Laser Confocal Microscopy: Challenging the Limits of Measuring Surface Roughness

3D imaging and ultra-high resolution are ideal for the inspection of materials with defined surface finishes and textures.

BY ROBERT BELLINGER, OLYMPUS SCIENTIFIC SOLUTIONS AMERICAS INC.

When evaluating the surface of a component, surface roughness can be assessed by eye or rubbed with a fingertip. Common expressions include "shiny," "lusterless and rough," "like oxidized silver" or "like a mirror." The differences indicated by these terms are caused by the variations in the irregularities of the component's surface.

The level of shininess — or asperity — of a surface is an important characteristic and can be quantified. Surface irregularities may be intentionally created by machining, but they can also be created by factors such as tool wobbling caused by motor vibration during machining, the quality of a tool edge or the nature of a machined material.

The form and size of irregularities vary, and are superimposed in multiple layers. Differences in these irregularities impact the quality and function of the surface. Irregularities affect the performance of the end product in aspects such as friction, durability, operating noise, energy consumption and airtightness.

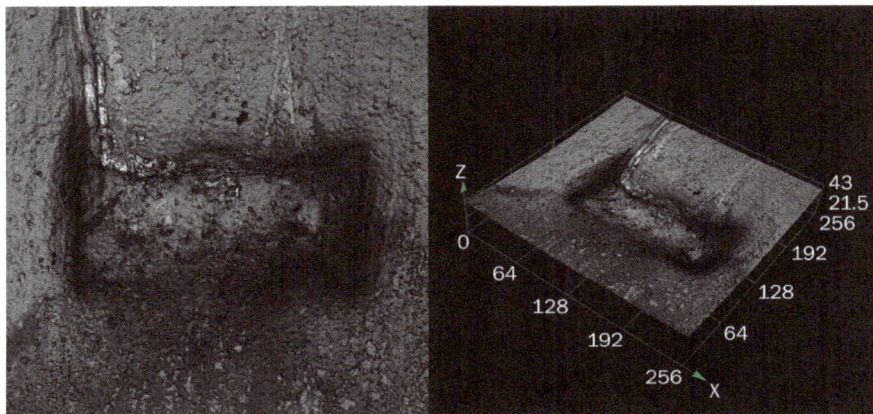

Figure 1. 2D image (left) and 3D image (right) of printed circuit board connector created using a laser confocal microscope. Images courtesy of Olympus Scientific Solutions Americas Inc.

Reasons for measuring surface roughness

The shape and size of irregularities on a machined surface have a major impact on the quality and performance of that surface. The quantification and management of fine surface irregularities is necessary to maintain high product performance.

Quantifying surface irregularities means assessing them by height, depth and interval. They are then analyzed by a predetermined method and calculated per industrial quantities standards. The form and size of surface irregularities and the way the finished product will be used determine if the surface roughness acts in a favorable or an unfavorable way. For example, surfaces that will be painted should be easy for paint to adhere to, and drive surfaces should rotate easily and resist wear. It is important to manage surface roughness so that it is suitable for the component in terms of quality and performance.

Many parameters have been established regarding the measurement and assessment of surface roughness. As machining technologies progress and higher-quality products are required, the performance of digital instruments, in particular laser confocal microscopes, continues to improve. The surface roughness of more diverse surfaces, including painted plastics, metal mating surfaces, solar panels and circuit boards, can now be assessed.

Two measuring methodologies

Surface roughness measurement methods include linear roughness measurement (profile method type), which involves use of a single line on the sample surface, and areal roughness measurement (areal method type), which acquires an area of the surface. Linear roughness measurement had been the industry standard for many years, but expectations for areal roughness measurement, which gathers a larger surface sampling area, have risen in recent years due to advances in laser confocal microscopy.

With linear roughness measurement (profile method type), the degree of roughness in the surface is measured along an arbitrary straight line. Long, continuous dimensions are measured and a contact stylus is commonly used to perform

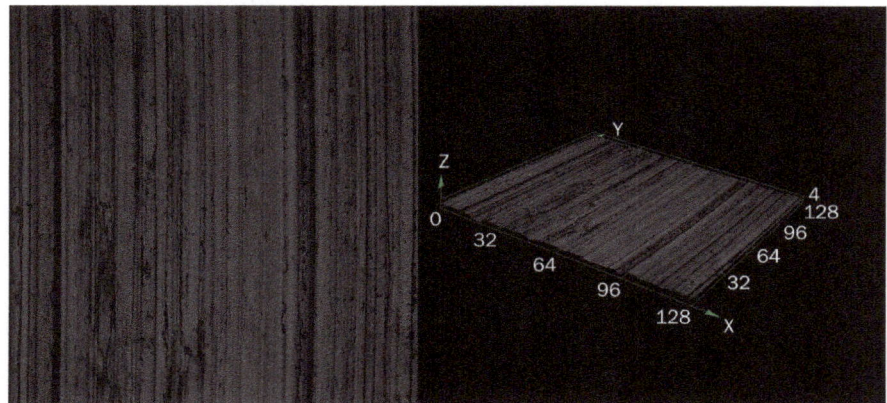

Figure 2. 2D image (left) and 3D image (right) of polished metal surface created using a laser confocal microscope.

roughness measurement. Linear roughness measurement is compliant with ISO and other national standards.

Laser scanning for areal roughness measurements

With areal roughness measurements (areal method type), the degree of roughness in the surface is measured over an arbitrary rectangular range. Areal roughness measurement provides a larger sampling area of the surface, providing a more accurate depiction of the state of the surface. A laser scanner is commonly used to perform areal roughness measurement.

Contact and noncontact measuring instruments

The instruments used for measuring surface roughness can be broadly divided into two types: contact and noncontact.

With the contact type, the tip of a stylus directly touches the surface of the sample (Figure 3). As the stylus traces across the sample, it rises and falls together with the roughness on the sample surface. This movement in the stylus is picked up and used to measure surface roughness. The stylus moves closely with the sample surface, so data is highly reliable.

The leading noncontact method involves light. Light emitted from an instrument such as a laser confocal microscope, focus detection system or interferometer is reflected and read, measuring without touching the sample (Figure 4). As they are noncontact, these systems never harm the sample and can even measure soft or viscous materials.

Noncontact observations in 3D

With laser confocal microscopy, noncontact 3D observations and measurements of surface features at submicron resolutions are easy to produce. Measurements of this nature are extremely important when analyzing and testing the surfaces of mating metals, especially within the areas of aerospace, satellite design, automobile production and military equipment production.

For example, if an engine component is going to be routinely subject to high RPMs, surface analysis at extremely high resolutions is imperative to ensure safety, reliability and performance. Additional benefits of laser confocal microscopy include fast image acquisition and high-resolution microscope images over a wide area, which can help save valuable testing and production time.

The tip radius of a general contact-type stylus is about 2 to 10 µm, which causes the roughness data to be "filtered" by the size of the stylus. In contrast, the radius of a laser spot from a laser confocal microscope is only 0.2 µm, so it can measure surface roughness that a contact-type stylus cannot enter (Figure 5).

High accuracy of laser confocal microscopes

With a contact-type instrument, it is extremely difficult to measure narrow

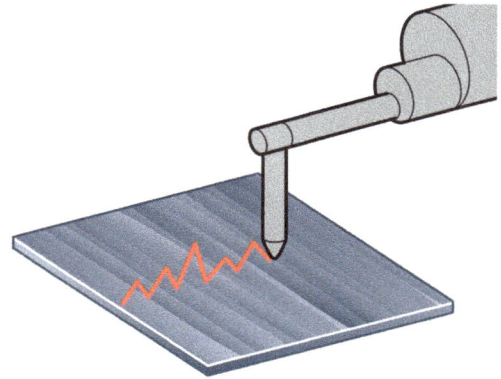

Figure 3. Linear roughness measurement performed with a contact stylus.

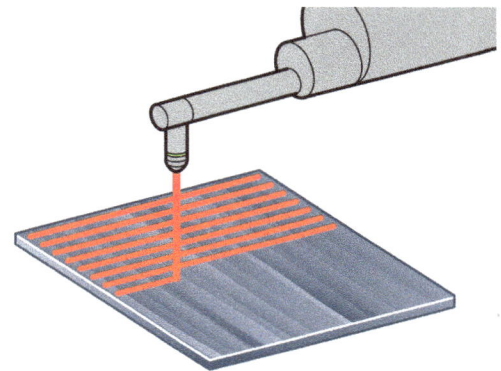

Figure 4. Areal roughness measurement performed with a laser scanner.

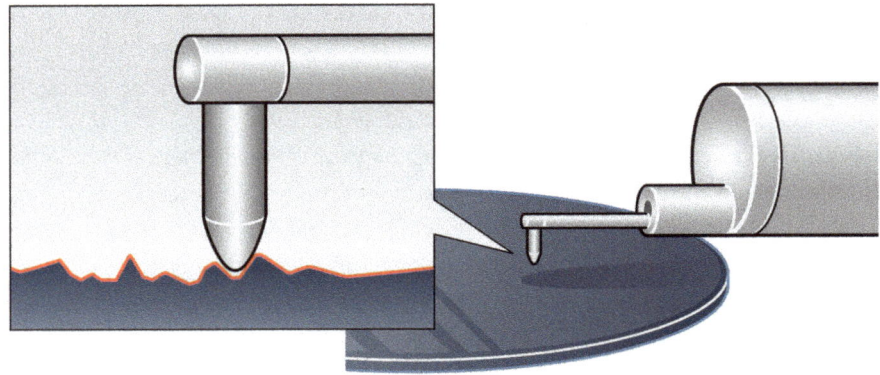

Figure 5. Contact-type roughness instrument.

areas such as fine wires (Figure 8). With a laser confocal microscope, however, positioning can be determined accurately, and it is easy to perform areal roughness measurement for a small target area (Figure 7).

The use of laser confocal microscopes for measurement is largely divided into two categories: horizontal measurement using an intensity image at a high resolution, and 3D measurement using a height image.

During horizontal measurement using an intensity image, the most important determining factor of measurement accuracy is control of the oscillation angle of the scanning mechanism. Many of today's newest laser confocal microscopes are periodically calibrated using a standard sample to ensure stable measurement for long periods of time. (The galvanometer mirror often used in the scan optical system uses a coil for position detection, so it takes some time to stabilize.) As the intensity dramatically changes around the focal position in the confocal optical system, focusing is one of the factors that affects the repeatability of a measurement result. When determining the line width of a pattern on a separated sample surface with a laser confocal microscope, it is desirable to use the relatively faster X-axis direction, which is not easily affected by vibration and other disturbances

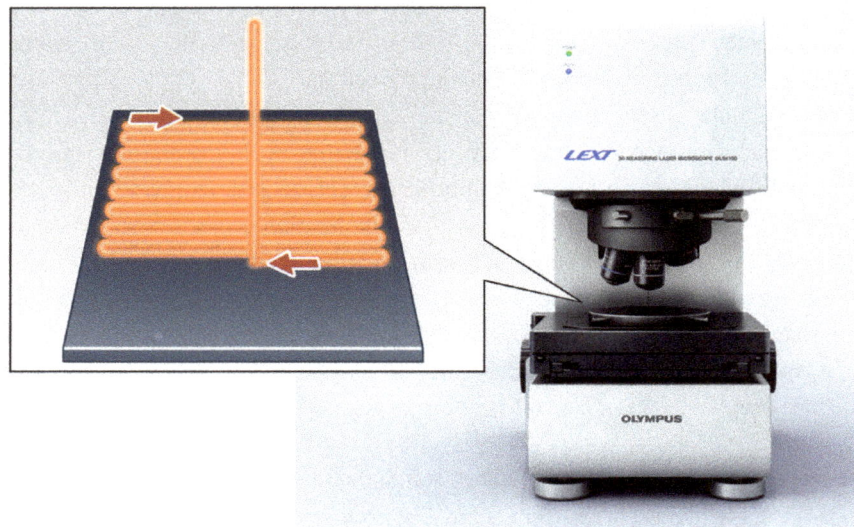

Figure 6. Laser confocal microscope (focus detection system).

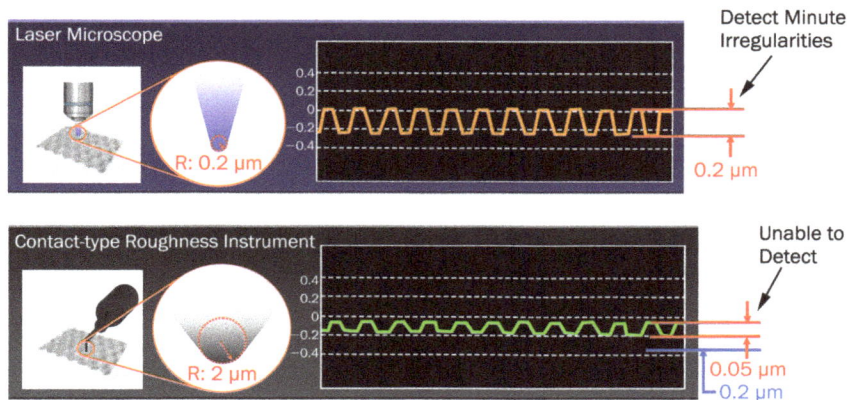

Figure 7. Laser confocal microscope vs. contact-type roughness instrument.

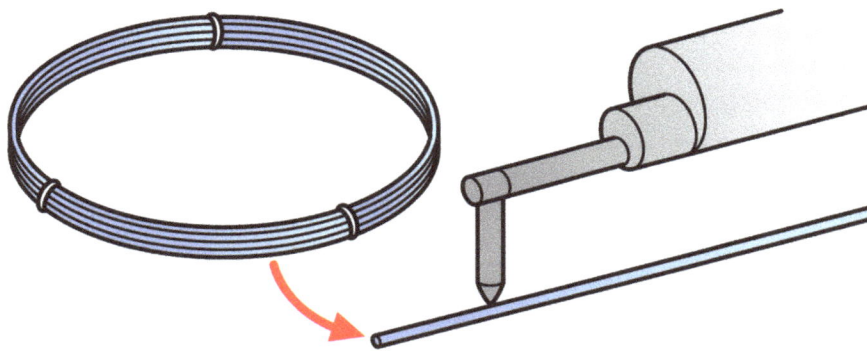

Figure 8. It is extraordinarily difficult to lower a stylus onto a wire surface only tens of microns across.

for measurement (especially because the X and Y axes have different speeds). A microelectromechanical systems (MEMS) mirror is used in the X direction to improve speed and accuracy.

Measuring in three dimensions

The driving factor for accuracy in 3D measurement is the Z-drive mechanism, which can move the objective lens and sample relative to each other. The method of driving the revolving nosepiece to which the objective lens is attached in the Z direction or the method of driving the XY stage on which the sample is positioned in the Z direction are the possible implementations of a Z-drive mechanism.

If a laser confocal microscope is used to measure the height of a sample, the highest intensity at each pixel must be found by moving in the Z direction. Therefore, the travel mechanism must be accurately moved with a resolution of about 10 nm at the highest magnification. A highly accurate linear guide and feeding screw are used together and a pulse motor or other device is used in most cases to drive in increments of a few millimeters, usually with a resolution accuracy of a few nanometers. A confocal optical system and a highly accurate Z-axis driving mechanism allow a laser confocal microscope to be used with the highest magnification objective lens and to resolve data down to several nanometers in the Z direction.

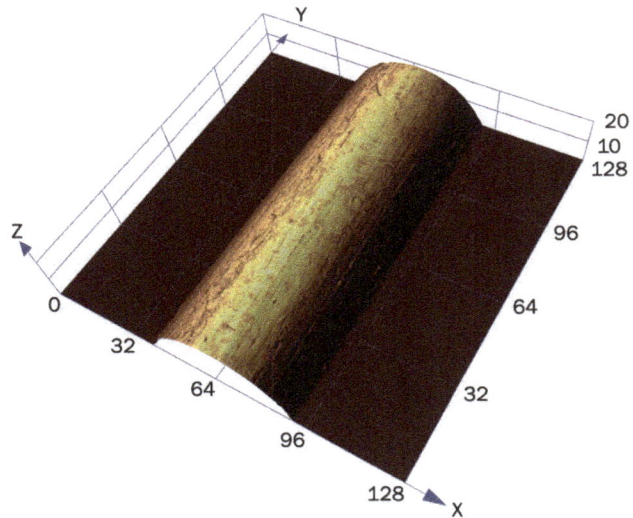

Figure 9. Observation image from a laser confocal microscope. Extremely fine wire (φ50 μm), objective lens magnification 100×.

Dental implants all roughed up

3D surface measurement using a height image is critical within a number of industries including the manufacture of medical implants and devices. In the case of dental implants, the surface roughness of the root portion of the implant is important because roughness increases the overall surface area which, in turn, increases the implant's stability. The metal portions of dental implants often undergo multiple processing stages to increase roughness including blasting, acid etching, anodic oxidation and polishing. The optimum surface roughness for dental implants is between 1 and 10 μm. In Japan, dental implant manufacturers are required to verify the measurements of surface roughness using a laser confocal microscope. Failure to have appropriate roughness can lead to poor adhesion of an implant to a patient's mouth.

Many of today's advanced laser confocal microscopes offer high levels of accuracy and precision in measuring the roughness of dental implant surfaces. The Olympus LEXT OLS4100, for instance, can measure steep angles so that shapes of varying geometry can be accurately and precisely measured (Figure 10). Utilizing a long working distance objective lens, the microscope allows more separation

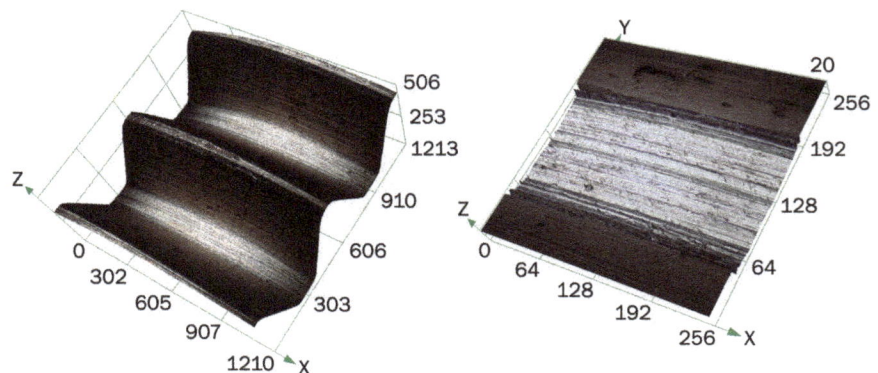

Figure 10. Dental implant surface measurements made using the Olympus LEXT OLS4100.

between the sample surface and the microscope lens so that larger objects, such as dental implants, can be accurately inspected.

Today's newest laser confocal microscopes deliver the ability to make 3D observations with ultra-high-resolution measurements and a high pixel density. Varying objectives give the user the flexibility of having a working distance capable of accommodating larger objects. High inclination sensitivity provides the ability to make accurate measurements of complex and steep-sided irregularities.

Widely used in quality control, research and development across an array of industries and applications, laser confocal microscopes have set new standards in 3D surface roughness measurement. Today, as demand grows for increased measurement precision and wider observation applicability, these instruments have continued to evolve to facilitate faster, easier measurement and higher-quality imaging. Benefits include fast noncontact, nondestructive measurement; accurate measurement of submicron distances across the XY axes; superior Z-axis measurement; wide sample ranges; high-angle measurement capabilities; performance guarantees; a wide range of measurement types; realistic surface reproduction; and crystal-clear 3D images.

Meet the author

Robert Bellinger is product applications manager at Olympus Scientific Solutions Americas Inc. He provides application support for Olympus industrial microscope systems in the U.S., Canada and Latin America.

Fluorescence Spectroscopy: A Powerful Method for Surface Analysis

Time-integrated laser-induced fluorescence spectroscopy is one of the most sensitive techniques based on the interaction between light and matter. It works on the basis of registering direct radiation emitted by the substances to be detected, and offers exceptional detection sensitivity and spatial resolution.

BY DR. JENS BUBLITZ, KIENZLE PROZESSANALYTIK GMBH

The analysis of technical surfaces is crucial for the optimization of lubrication, coating or cleaning processes, especially with regard to achieving and assuring better quality of final products. Frequently, the essential requirements for an additional, specific production step cannot be controlled sufficiently, since most laboratory methods provide only limited and randomly appropriate information. An efficient inline and online process analysis can, therefore, contribute significantly to the prevention of quality defects. The reduction in production losses saves resources and essentially increases the added value of a process. In principle, the required surface analysis can be divided into two different application areas: first, the monitoring and evaluation of functional coatings, and second, the analysis of surface cleanliness.

Common methods — chemical analysis, thermal treatment of product samples with a subsequent gravimetric evaluation, or CO_2 (total organic carbon, or TOC) gas analysis — are based on taking product samples. Plasma or mass spectroscopy are vacuum-based methods, where an invasive interaction with the respective surface takes place. Comparatively, optical methods have the big advantage of an analysis without any contact to the surface to be investigated. However, most currently used methods, such as IR-, Raman- or diffuse-reflectance spectroscopy,

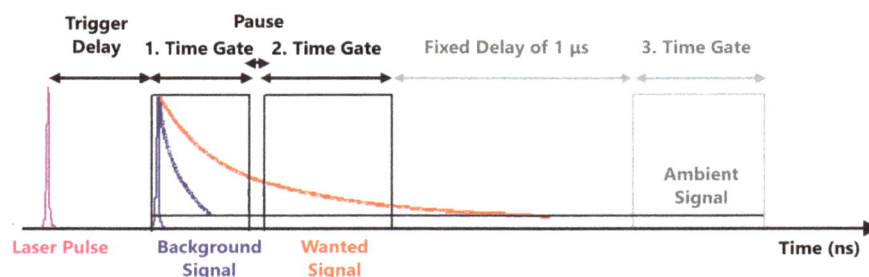

Figure 1. Principle of time-integrated fluorescence detection.

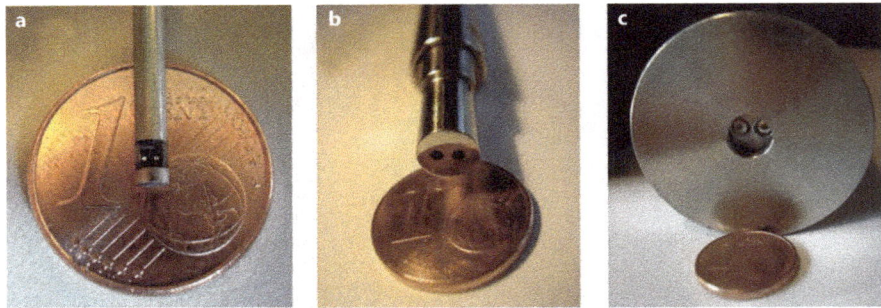

as well as the ellipsometry, are dependent on constant reflection properties of the examined product surface. Furthermore, to determine a surface amount or layer thickness derived from transmission or absorption behavior, an initial intensity is always needed as a reference value.

The special time-integrated laser-induced fluorescence spectroscopy, or LIF(t), technique is a process-analytic, contact-free procedure that allows a nondestructive analysis of the surface cleanliness directly in production processes. The method can also be applied for qualitative and quantitative monitoring of lubrications or other applied functional coatings.

Physical and technical principle

The basic principle of laser-induced fluorescence spectroscopy is the absorption of the exciting laser radiation through the substance to be detected. In this process, an interaction of laser radiation with fluorophore regions in the electron structures of the target molecule takes place. The absorbed energy is re-emitted by the molecules as a new light, the so-called fluorescence, in times on a nanosecond scale (10^{-9} s). This process is one of the most efficient interactions between light and matter, and it occurs in dependency of the substances involved and substrate media, with measurably different speeds. A statistical registration of the single photons here offers, in comparison to other spectroscopic methods, a detection of very low quantities of material with particularly high sensitivity.

Laser emission in the UV spectral range (e.g., 266 or 355 nm) might lead to a fluorescence excitation of the molecules that should be detected (wanted signal), as well as a possible background signal caused by organic matter on the surface below the analytical target.

Registering only the spectral intensity distribution of the fluorescence doesn't necessarily lead to a significant separation of the substance spectra. Therefore, a time-integrating approach is included in the procedure to observe the decay times of fluorescence signals in a suitable wavelength range. After each single laser pulse excitation, lasting about 0.5 ns, the time decay of the fluorescence radiation is registered in two appropriately positioned windows. Here, the time-integrated intensity is measured as values I1 and I2 in order to separate the wanted signal from the background, as indicated in Figure 1. In a third time window, positioned about 1 µs after the second one, the ambient light intensity I3 is registered and used to correct the values of I1 and I2.

Optical excitation is performed by a specially designed, passively Q-switched UV microchip laser from Teem Photonics SA with a repetition frequency in the range of 11 kHz. Single laser pulses with an emission wavelength of 266 or 355 nm and an average optical output power below 2 mW are generated. These are transferred via a quartz fiber bundle of up to 20 m in length and a probe head directly into the process. Here, a typical output power of 250 µW is used.

With a second quartz fiber bundle, the transmission of fluorescence signals to the detector is arranged. Therefore, the corresponding detection head contains no active components, and is suitable to be used in harsh and demanding environmental conditions. If there is a risk of contamination of the fiber optics, automated pollution prevention is ensured by using a compressed air or gas flushing, or appropriate ultrasound techniques. Figure 2 shows examples of three different configurations for various applications of inline surface analysis.

By choosing the configuration and positioning of the probe (distance and angle) with respect to the surface of investigation, the spot size of the effective detection area can be adapted to the process requirements.

For a very sensitive detection of single-fluorescence light events (photons), a

The basic principle of laser-induced fluorescence spectroscopy is the absorption of the exciting laser radiation through the substance to be detected.

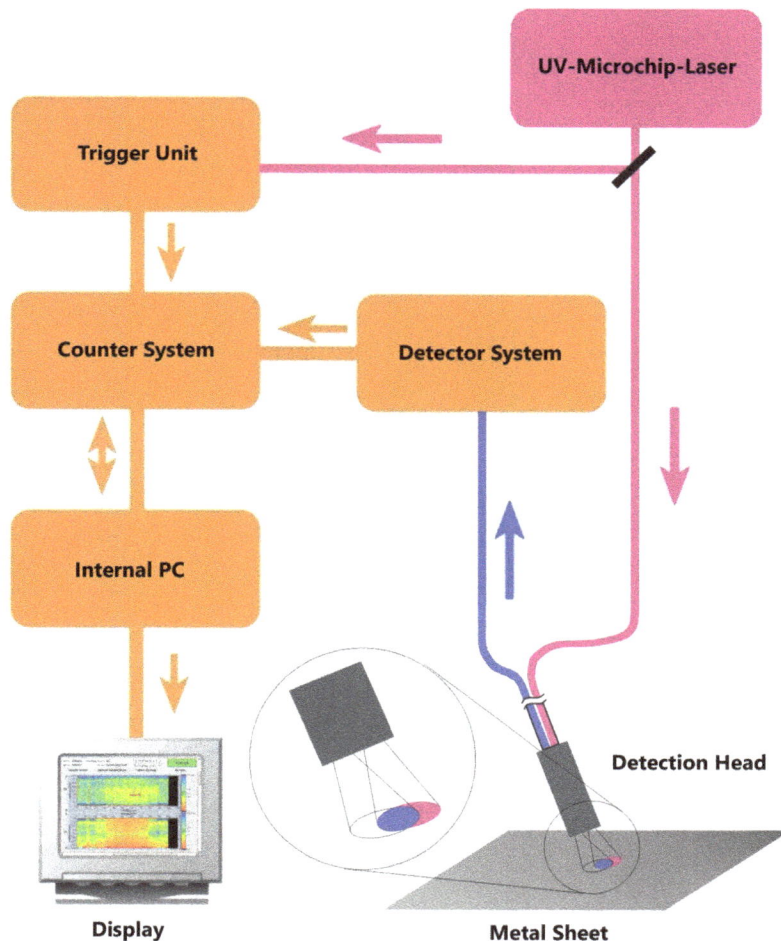

Figure 3. Schematic setup of the time-integrated fluorescence detection system.

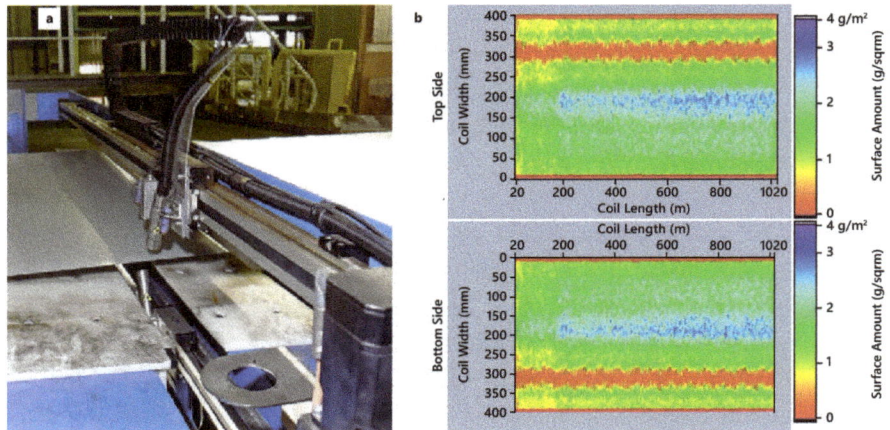

Figure 4. Fluorescence detection system for a simultaneous inline and online inspection of the top and bottom sides of lubricated coil material: (a) installation situation and (b) typical presentation of the results in a color gradient graph.

photomultiplier tube is used. An application-specific wavelength configuration is provided by a special combination of optical filters in front of the detector. Using a statistical single-photon counting method, the detector pulses are evaluated according to the time-integrated method. Via a calibration function, the respective measured quantity is calculated and displayed depending on the application with a high sampling rate of every 10 ms or even every second. Since each single-fluorescence measurement takes place in the nanosecond timescale, the results of surface measurements on moving parts or sheets do not depend on the process speed, which could be more than 2000 m/min. Figure 3 demonstrates schematically the technical implementation of the time-integrated laser-induced fluorescence analysis.

Industrial applications

A versatile field of application of coating analysis is the specific detection of lubricants on coil material (single boards), and on prefabricated parts in the steel and aluminum industries. Coil material is customer-specific and lubricated with different kinds of prelubes for the automotive industry. Here, the laser-induced fluorescence detection technique is able to monitor inline and online deviations from the specified layer thickness in a typical range of up to 3 g/m^2, and in particular to find possible dry areas with size greater than 15 mm at line speeds of up to 600 m/min. Figure 4a shows a typical installation of a system for simultaneous inspection of the upper and lower surfaces of coil material with a width of up to 2 m. It was integrated in an existing production line.

For the measurements, two single detection heads with compressed-air flushing (Figure 2c) are synchronously moved with a travel speed of 0.5 m/s across the surface of the metal strip, and the current local layer thickness is determined quantitatively with a sampling rate of 10 ms. Figure 4b demonstrates typical results for the top and bottom surface, each in separate color gradient plots. The green indicates an optimum accordance with the predefined amount of prelube for the lubrication; the blue color indicates an over-oiling. The red is a warning for dry areas, which in most cases are not observable by a direct visual inspection.

a

b

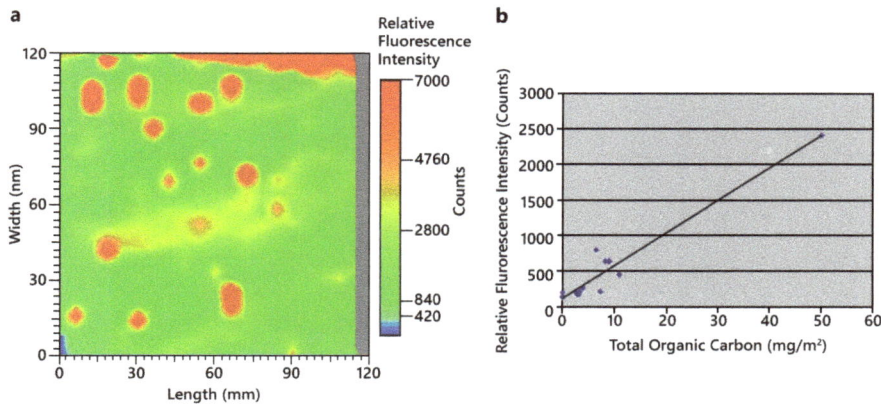

Figure 5. Typical results for detection of rolling-oil residues on aluminum sheets: (a) color gradient plot of measurement with a two-axis translator and (b) correlation of the fluorescence data to the results of a surface carbon analyzer.

Beyond lubrication control, there is a variety of other applications in the field of metal processing, where the inline monitoring of special functional layers might increase the process efficiency, or lead to better, more stable and reliable product quality. Some examples include passivation coatings, lacquers, films, polymers, primers and adhesives.

In the metal working industry, large amounts of production additives, such as cooling lubricants, are used to ensure high process reliability in forming or machining production. Residuals of these coatings remain as contaminants on the workpiece surface and will be unavoidably carried over into subsequent process steps. Special manufacturing processes are very sensitive to the surface cleanliness of the working material, such as annealing, bonding, coating, painting, etc. Therefore, in most applications, the workpiece is cleaned prior to further processing. The continuous monitoring and evaluation of technical surfaces concerning residues of process additives and cleaning media is another important application of the fluorescence spectroscopic surface analysis.

In the case of production that uses the rolling process, aluminum, steel or copper sheets are annealed in order to remove remaining rolling oil on the surface. Commonly, customers of an aluminum or copper plant require a leftover of rolling oils and other production media of not more than 5 mg/m^2 on the surfaces of their final products.

Figure 5 shows the results of such a determination of rolling-oil residues on aluminum sheets. By using a two-axis translator unit, a complete area screening of sample sheets in the size of 100 × 100 mm with a spatial resolution of 2 mm is possible in order to determine the distribution of locally varying degrees of contamination (Figure 5a); a comparison of the fluorescence data averaged over the entire sample surface is plotted in Figure 5b. The results were determined subsequent to the fluorescence measurements with a standard surface carbon analyzer, based on the determination of the TOC.

Meet the author

Dr. Jens Bublitz is manager of R&D at KIENZLE Prozessanalytik GmbH in Flensburg, Germany.

Lasers and Optics

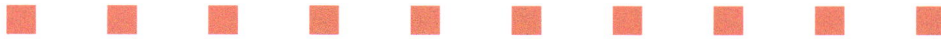

Lasers:
Understanding the Basics

Although lasers range from quantum-dot to football-field size and utilize materials from free electrons to solids, the underlying operating principles are always the same. This article provides the basic information about how and why lasers work.

COHERENT INC.

Over 50 years have passed since the first demonstration of a laser in 1960. After the initial spark of interest, lasers were for a while categorized as "a solution waiting for a problem," but bit by bit, the range of their applications has expanded to encompass fields as diverse as DNA sequencing, consumer electronics manufacturing or freezing the motion of electrons around atoms. Most of these applications simply would not have been possible without lasers. To grasp the relevance of lasers in physics, it is enough to note that no other man-made sources can generate pulses (of any type) as short as laser pulses — now below to 10^{-16} s — or tools to measure absolute frequencies with an accuracy of $\sim 10^{-15}$! Industrial manufacturing, microelectronics, and biomedical and instrumentation applications serviced by lasers are incredibly diverse and rely on unique capabilities, like producing features below the limit of light diffraction, modifying materials in their bulk while leaving the surface unaffected, or trapping and moving individual particles in mid-air.

All light sources convert input energy into light. In the case of the laser, the input, or pump, energy can take many forms, the two most common being optical and electrical. For optical pumping, the energy source may be a lamp or, more

Figure 1. Spontaneous emission is a random process, whereas stimulated emission produces photons with identical properties. Images courtesy of Coherent.

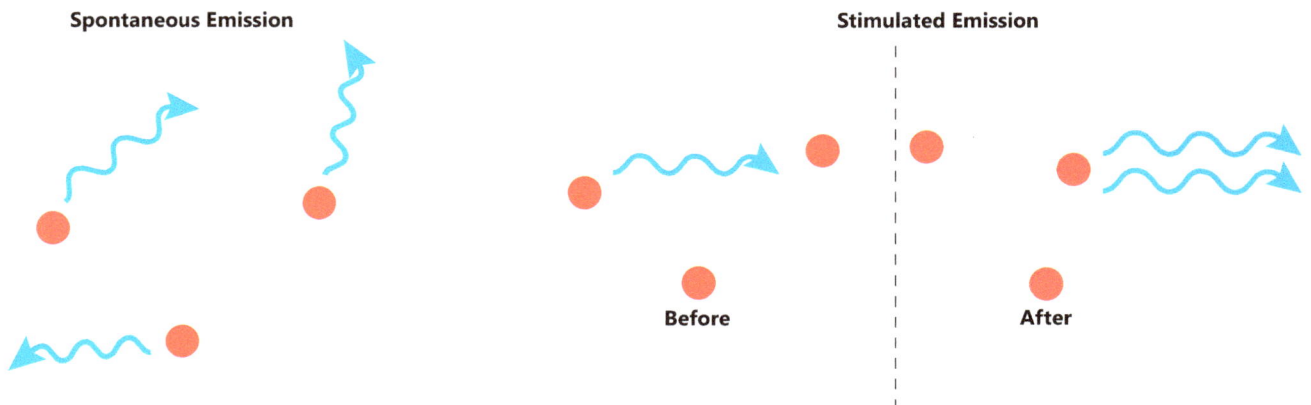

Spontaneous Emission

Stimulated Emission

Before

After

commonly, another laser. Electrical pumping can be via a DC current (as in laser diodes), an electrical discharge (noble gas lasers and excimer lasers) or a radio-frequency discharge (some CO_2 lasers).

In a conventional (incoherent) light source like a lightbulb, an LED or a star, each atom excited by input pump energy randomly emits a single photon according to a given statistical probability. This produces radiation in all directions with a spread of wavelengths and no interrelationships among individual photons. This is called spontaneous emission.

Einstein predicted that excited atoms also could convert stored energy into light by a process called stimulated emission. This process typically starts with an excited atom first producing a photon by spontaneous emission. When this photon reaches another excited atom, the interaction stimulates that atom to emit a second photon (Figure 1). This process has two important characteristics. First, it is multiplicative — one photon becomes two. If these two photons interact with two other excited atoms, this will yield a total of four photons, and so forth. Second and most importantly, these two photons have identical properties: wavelength, direction, phase and polarization. This ability to "amplify" light in the presence of a sufficient number of excited atoms leads to "optical gain" that is the basis of the laser operation and justifies its acronym of light amplification (by) stimulated emission (of) radiation. A wide range of solid, liquid and gas-phase materials have been discovered that exhibit gain under appropriate pumping conditions.

The laser cavity

The laser cavity, or resonator, is at the heart of the system. A single transit through a collection of excited atoms or molecules is sufficient to initiate laser action in some high-gain devices such as excimer lasers; however, for most lasers, it is necessary to further enhance the gain with multiple passes through the laser medium. This is implemented along an optical axis defined by a set of cavity mirrors that produce feedback (Figure 2). The lasing medium (a crystal, a semiconductor or gas enclosed in an appropriate confinement structure) is placed along the optical axis of the resonator. This unique axis with very high optical gain becomes also the direction of propagation of the laser beam. A somewhat different example of a uniquely long (and flexible!) gain axis is the fiber laser.

The simplest cavity is defined by two mirrors facing each other — a total reflec-

Figure 2. In the prototypical gas laser, the gain medium has a long, thin cylindrical shape. The cavity is defined by two mirrors. One is partially reflecting and allows the output beam to escape.

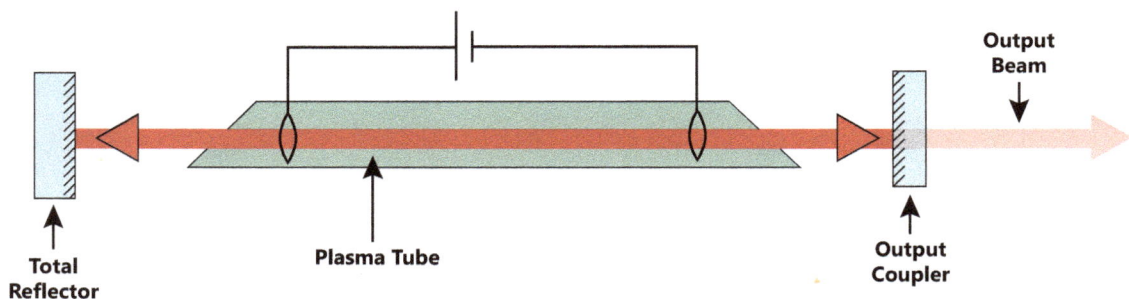

Total Reflector

Plasma Tube

Output Coupler

Output Beam

tor and a partial reflector whose reflectance can vary between 30 and close to 100 percent. Light bounces back and forth between these mirrors, gaining intensity with each pass through the gain medium. Photons that are spontaneously emitted in directions other than the axis are simply lost and do not contribute to the laser operation. As laser light is amplified, some of the light escapes the cavity, or oscillator, through the partial reflector (output coupler); however, at equilibrium (the so-called "steady state" or "continuous wave"), these "optical losses" are perfectly compensated by the optical gain experienced in the successive round-trip of the photons inside the cavity. The output of the laser is exactly the part of the beam transmitted by the output coupler. In an ideal laser, all the photons in the output beam are identical, resulting in perfect directionality and monochromaticity. This determines the unique coherence and brightness of a laser source.

Monochromaticity — A photon's energy determines its wavelength through the relationship $E = hc/\lambda$, where h is Planck's constant, c is the speed of light and λ is wavelength. An ideal laser would emit all photons with exactly the same energy, and thus the same wavelength, and it would be perfectly monochromatic. Many applications are dependent on monochromaticity. For example, in telecommunications, several lasers at slightly offset wavelengths can transmit in parallel streams of pulses down the same optical fiber without crosstalk. Real lasers are not perfectly monochromatic because several broadening mechanisms widen the frequency (and energy) of the emitted photons. For example, free-running YAG lasers can have linewidths of hundreds of gigahertz, while stabilized diode-pumped YAG lasers can have a linewidth <1 kHz. The best known of these broadening mechanisms is the Doppler broadening, determined by the distribution of speeds in the collection of atoms making up active gas mediums.

Coherence — Besides sharing the same wavelength, the photons that make up a laser beam are all in phase (Figure 3), or "coherent," resulting in an electric field that propagates with a uniform wavefront. The ideal representation is a plane wave that propagates with a flat wavefront along a given direction and where each plane perpendicular to this direction experiences the same electric and magnetic field amplitude and phase at a given time. When two waves with such characteristics

Figure 3. Laser light differs from conventional light in that all the lightwaves are in phase with each other.

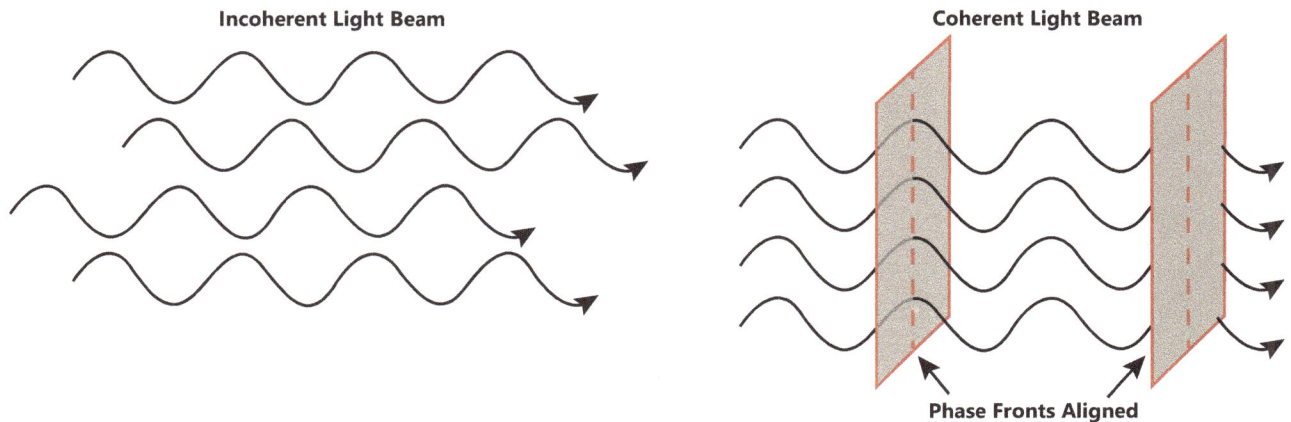

Incoherent Light Beam

Coherent Light Beam

Phase Fronts Aligned

interact, they create interference patterns, as in Young's experiment. Real laser beams somewhat deviate from this ideal behavior, but they are still the sources that best approximate an ideal coherent plane wave, and they enable a host of applications that rely on optical interference. For example, the surface of precision lenses and mirrors is measured using laser interferometers, and so are the minute variations in the interference patterns of miles-long interferometers used to chase and detect gravitational waves.

Brightness (or, more correctly, radiance) — The most strikingly visible difference between lasers and conventional light sources is that all the emitted light

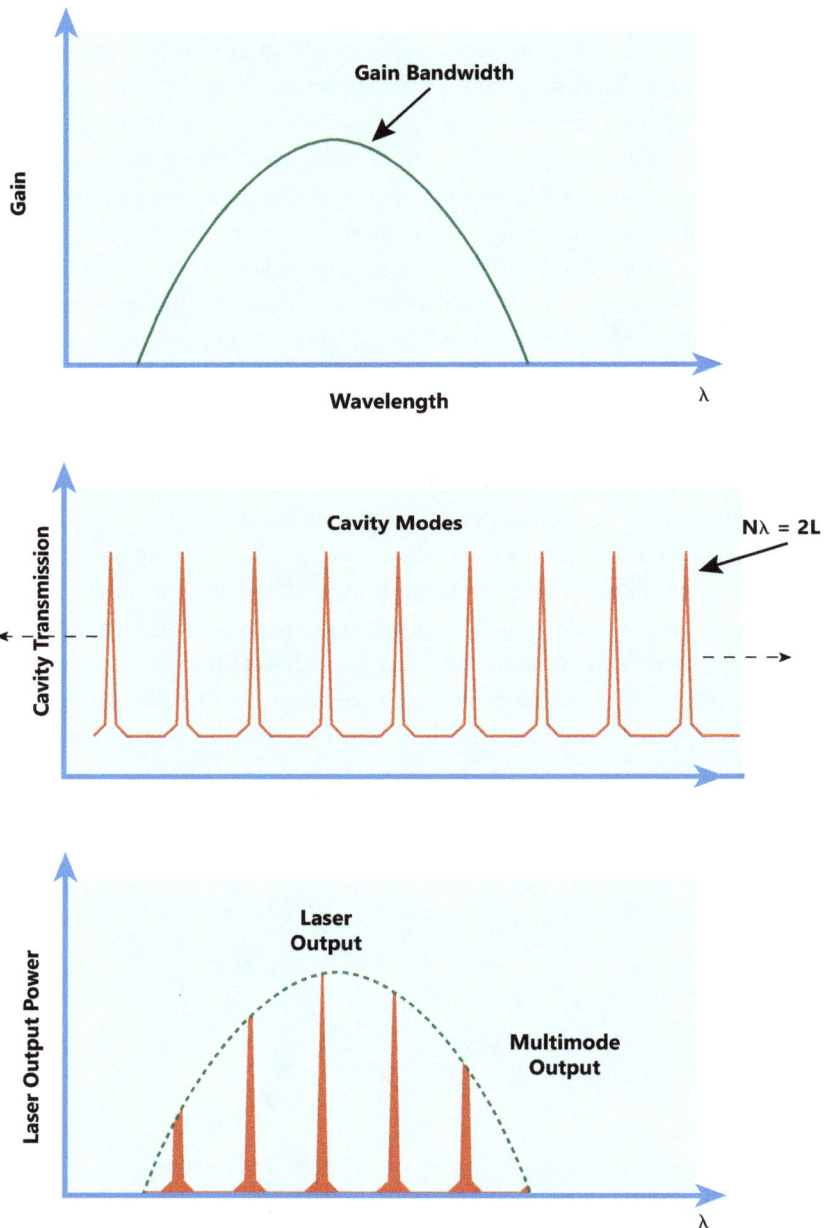

Figure 4. A resonant cavity supports only modes that meet the resonance condition, $N\lambda = 2 \times$ cavity length. The output of a CW laser is defined by the overlap of the gain bandwidth and these resonant cavity modes.

travels in the same direction as an intense beam. Radiance is defined as the amount of light leaving the source per unit of surface area and unit of solid angle. A star like the sun emits a large amount of radiation from a unit of surface area, but this is emitted in many different directions. On the contrary, a laser beam is highly directional, with the result that its brightness is much more intense than the sun's as experienced on the Earth's surface. For this reason, just 5 mW of power from a laser pointer is more "blinding" (and dangerous) for the eye than direct sunlight.

Because of its high radiance, a laser beam can be projected over great distances or focused to a very small spot. Well-designed lasers produce a beam of light that will expand ("diverge") only by the minimum amount prescribed by the laws of diffraction. For example, diffraction imposes that the minimum spot that can be produced by a laser beam is equal to about its wavelength.

Continuous-wave lasers

Lasers can be divided into three main categories: continuous wave (CW), pulsed and ultrafast.

As their name suggests, continuous-wave lasers produce a continuous, uninterrupted beam of light, ideally with a very stable output power. The exact wavelength(s) or line(s) at which this occurs is determined by the characteristics of the laser medium. For example, CO_2 molecules readily lase at 10.6 µm, while neodymium-based crystals (like YAG or vanadate) produce wavelengths in the range between 1047 and 1064 nm. Each laser wavelength is associated with a linewidth, which depends on several factors: the gain bandwidth of the lasing medium and the design of the optical resonator, which may include elements to purposely narrow the linewidth, like filters or etalons.

If a laser can simultaneously produce different lines, the first step in determining the operating wavelength is to use cavity mirrors that are highly reflective only at the desired wavelength. The low reflectivity of the mirrors at all the other lines will prevent these from reaching the threshold for laser action. However, even a single laser line actually covers a range of wavelengths. For example, laser diodes produce light over a wavelength range of several nanometers corresponding to their "gain bandwidth."

The specific wavelengths of the output beam within this gain bandwidth are determined by the longitudinal modes of the cavity. Figure 4 shows the behavior of a two-mirror cavity, the most basic design. To sustain gain as light travels back and forth between the mirrors, the waves must remain in phase and "reproduce" their wave pattern, which means that the cavity round-trip distance must be an exact multiple of the wavelength:

$$N\lambda = 2 \times \text{Cavity Length}$$

where λ is the laser wavelength and N is an integer called the mode number. This is usually a very large integer, since the wavelength of light is so much smaller than a typical cavity length. In a high-power laser diode, for example, the IR output wavelength is 0.808 µm, yet the cavity length may be 1 mm, so that even in a very small laser resonator, N is ~2500. Wavelengths that satisfy this resonance equation are called longitudinal cavity modes. The actual output wavelengths of the laser will correspond to the cavity modes that fall within the gain bandwidth, as shown in Figure 4 (bottom). This regime is called multilongitudinal-mode operation. Using the example of the high-power laser diode, the spacing between adjacent longitudinal modes is:

$$\sim\!150 \text{ GHz (equivalent to } \sim\!0.3 \text{ nm in wavelength difference)}$$

If the laser diode operates on a 3-nm gain line, about 10 longitudinal modes, spanning 3 nm, will be able to oscillate. The resonator design also controls the so-called transverse modes, responsible for the intensity distribution on the plane perpendicular to the beam direction. The ideal laser beam has a radially symmetric cross section: The intensity is greatest in the center and tails off at the edges, following a Gaussian profile. This is called the TEM_{00} or fundamental output mode. Lasers can produce also many other TEM modes, a few of which are shown in Figure 5. Usually a round aperture placed inside the cavity is used to force the laser to operate in the fundamental mode. In multitransverse-mode operation, many modes are present at the same time, often resulting in a profile that appears to be Gaussian but in reality has degraded properties (higher divergence and lower radiance). Often the quality of a laser beam is specified using the M^2 (M squared) parameter. For example, a YAG laser operating exclusively in the TEM_{00} mode has $M^2 = 1$, while multimode laser diodes have an M^2 of hundreds. Different transverse modes also have slightly different frequencies; however, this difference is much smaller than the difference between adjacent longitudinal modes (~1 MHz compared with approximately hundreds of megahertz to hundreds of gigahertz).

A laser that produces multiple longitudinal modes has a limited coherence — different wavelengths cannot stay in phase over extended distances. Applications such as holography, which demand excellent coherence, benefit from using a single-longitudinal-mode laser. For some laser types with a narrow gain bandwidth, single-mode output is achieved with a very short resonant cavity; this makes the mode spacing larger than the gain bandwidth, and only one mode lases. Generally, though, a filtering element that preferentially passes only one mode is inserted into the cavity. The most common type of filter is called an etalon. Using a number of sophisticated design enhancements, it is possible to restrict the linewidth of a laser to less than 1 kHz, useful for scientific interferometric applications.

Some solid-state lasers have extremely broad bandwidths that extend to hundreds of nanometers. The most common example is the Ti:sapphire laser. Rather than being a disadvantage, this broad bandwidth enables the design of tunable and ultrafast (femtosecond and picosecond pulse width) lasers. Designing a tunable CW laser involves including an extra filtering element in the cavity — usually a birefringent (or Lyot) filter. A birefringent filter does two things: It narrows the bandwidth and, by rotating the filter, allows smooth tuning. This same type of filter is also used as a factory-set tool to lock the wavelength at a precise value, when broad-bandwidth lasers need to be preset at a specific application-dependent wavelength. This is typically the case with optically pumped semiconductor lasers (OPSLs) that can be set at the desired wavelength within their 5- to 10-nm operating range.

Most applications of CW lasers require that the power be as stable as possible over long time periods (hours or weeks), as well as over short time durations (microseconds), depending on the specific application. To ensure this stability also in the presence of varying environmental situations like temperature, vibration and the aging of the laser itself, microprocessor control loops are implemented. For example, a diode-pumped Nd laser will have servos to adjust temperature and output power of the pump diodes to maintain stable output power from the resonator. In addition, other servos may control the perfect alignment of the resonator mirrors.

Pulsed lasers

Some materials — like excited dimers (or "excimers") of a noble gas with a halogen, such as ArF and XeCl — sustain laser action for only a brief period of several nanoseconds. Other lasers, like Nd or Yb diode-pumped solid-state (DPSS) lasers, lend themselves to be operated both in CW or pulsed operation. Other lasers, like laser diodes or OPSLs, are not suitable at all for pulsed operations. Within this

Figure 5. Lasers can emit any number of transverse modes, of which the TEM_{00} usually is most desirable.

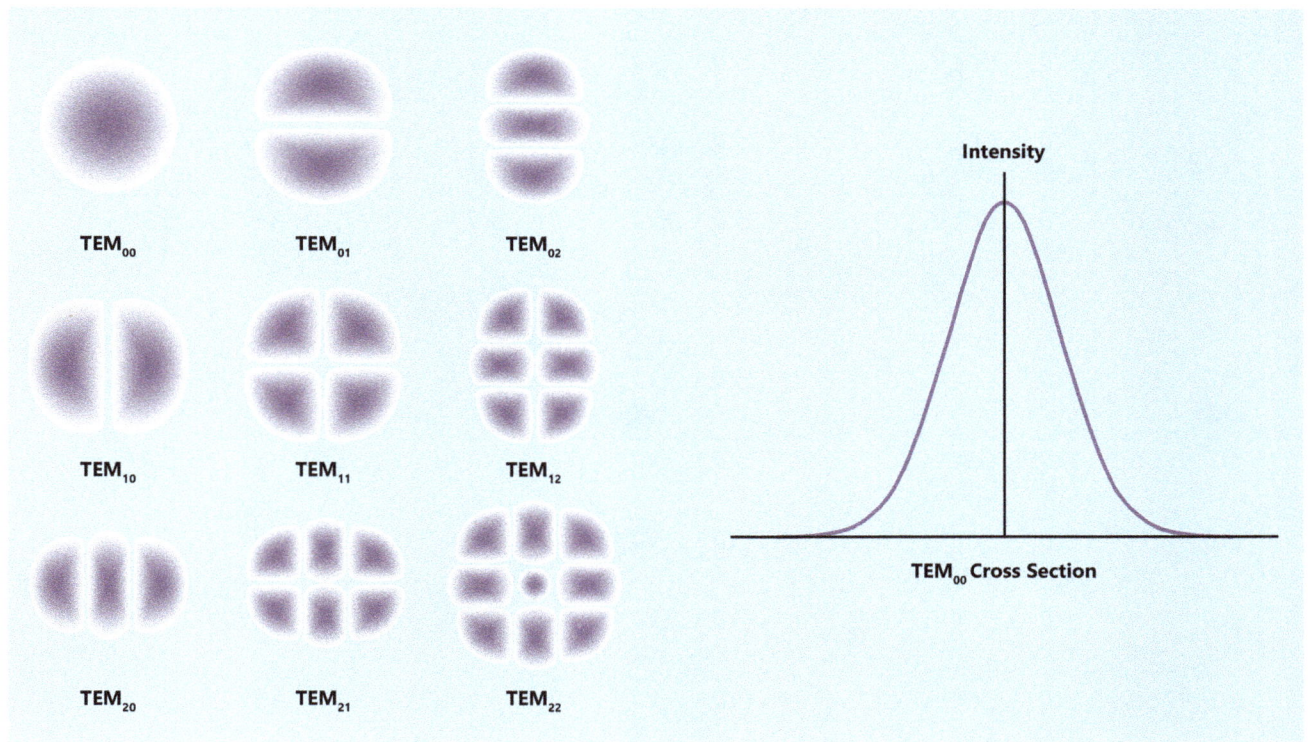

context, we define as "pulsed" laser devices that produce pulses of 0.5 to 500 ns. This regime is useful for time-resolved scientific experiments but especially for a vast range of manufacturing processes related to ablation or some other type of nonthermal materials modification. The most important characteristic of a nanosecond-pulsed laser is the capability to "store" and release energy very rapidly; i.e., on a nanosecond scale so that the laser output can achieve tens of kilowatts to megawatts of peak power. It is precisely this high peak power that enables the ablative processing of materials. In addition, the high peak power enables a number of so-called optical nonlinear processes; i.e., processes that rely on the interaction of more than one photon at a time with matter.

Operating a nanosecond-pulsed laser is substantially different from operating a CW laser. To build and produce each pulse, the light has time for very few round-trips in the laser cavity, and the simple two-mirror cavity based on a partly transmissive mirror described so far cannot produce these energetic and short pulses. The key to producing these energetic pulses is storing energy from the pump in the atoms or molecules of the lasing medium by preventing the laser gain and the amplification process. Then, when the stored energy is at its maximum, lasing action is rapidly enabled: The stored energy results in an extremely high laser gain (amplification) that takes place during only a few round-trips, during which a giant pulse builds up and gets coupled through the partly transmissive mirror. This regime is called Q-switched operation and can be conceptualized as a two-mirror cavity with an optical gate located between one of the mirrors and the laser medium (Figure 6). When the gate is closed and the laser medium is pumped, photons cannot circulate in the cavity, and the excitation of the atoms builds up; as soon as the gate is opened, photons start to build up via stimulated emission with

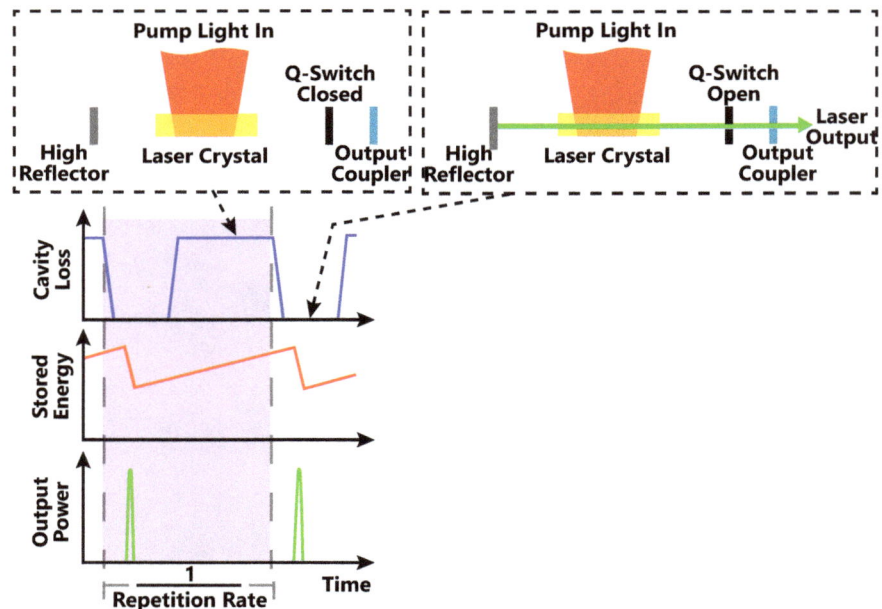

Figure 6. Schematic showing the operating principle of a Q-switch in a solid-state laser.

a very large gain at each round-trip; a fraction of them (~20-40%) get coupled by the partly transmissive mirror. The result is a pulse with a very sharp raising time and a slower falling time, with a typical duration of 1 to 200 ns. The pulse duration depends on several parameters: the type of gain medium and how much energy it can store, the cavity length, the repetition rate of the pulses and the pump energy, to mention the most important ones. Q-switched lasers commonly used in the industry can produce average powers up to tens or hundreds of watts and repetition rates as low as 10 Hz or as high as 200 kHz. Most industrial processes are in the kilohertz to tens-of-kilohertz regime.

The actual Q-switch device is an acousto-optical modulator or an electro-optical modulator (EOM). Both use crystals where an applied electric field produces some perturbation of the optical properties of the crystal. In the case of acousto-optical modulators, the applied electric field is a radio-frequency voltage that produces a high-frequency sound wave in the crystal. This sound wave diffracts the photons from the laser and prevents laser amplification. EOMs instead use an applied high voltage that modifies the crystal refractive index and alters the polarization of the incoming light; an appropriate combination of polarization-sensitive optics can be placed in the cavity to prevent light of altered polarization from circulating.

Other types of lasers, such as excimer lasers, do not require a Q-switch to produce nanosecond pulses but rather rely on a transient pump pulse. Excimer laser pulses are produced by exciting the noble gas/halogen mixture with a powerful and short electric discharge. Ti:sapphire lasers can also produce nanosecond pulses if they are pumped with a nanosecond pulse of green light produced by a frequency-doubled, Q-switched YAG laser. This method is called gain switching because the cavity gain rather than the cavity loss is directly changed.

Apart from a huge number of industrial applications, Q-switched lasers have important applications in scientific research. One is pumping of Ti:sapphire ultrafast amplifiers (described in the following section) by using the frequency-doubled (green) output of a Q-switched Nd:YAG or Nd:YLF at 1-10 kHz. Another one is using the YAG or YLF laser to produce energies per pulse in the joule range at 1-100 Hz. These lasers are often used with nonlinear optical generators that can produce tunable wavelengths in the UV, visible and IR region, enabling time- and wavelength-resolved studies. Nowadays most YAG or YLF lasers operating at >100 Hz are diode-pumped, while high-energy 10-Hz systems require pumping with a flashlamp because diodes are not suitable for producing high-energy output pulses.

For some scientific applications, it may be desirable to have a narrow-linewidth Q-switched laser. In certain cases, this can be accomplished using a combination of optical gratings and etalons; in other cases, the laser can be "seeded" with a low-power CW or Q-switched narrow-linewidth laser that is easier to control than the higher-power stage. This approach, called "injection seeding," uses a MOPA (master oscillator, power amplifier), conceptually splitting the linewidth selection and the high-power generation into two stages that are optimally designed for the two purposes.

Ultrafast lasers

Ultrafast lasers are generally defined as lasers that produce pulses in the range of 5 fs to 100 ps (1 femtosecond = 10^{-15} seconds). CW lasers can produce many longitudinal modes; if all these modes can be locked in phase (mode-locking regime), the resultant overlap of the optical field of all the modes will not be a standing wave but rather a distribution corresponding to very short pulses spaced by the cavity round trip. This mode-locked regime results in a single pulse traveling back and forth inside the cavity: Every time this pulse reaches the output coupler, the laser emits a part of this pulse. The pulse repetition rate is determined by the time it takes the pulse to make one trip around the cavity.

To understand the concept of locking together these modes, one can think about the various modes as having a "zero" (i.e., null electric field), all at a certain position (function of time) in the resonator. It turns out that the more modes that interfere, the shorter the pulse duration (Figure 7). Since larger lasing bandwidths support a larger number of oscillating modes, the pulse duration is inversely proportional to the bandwidth of the laser gain material. This explains why materials used for broadly tunable lasers produce also the shortest mode-locked pulses. The most popular ultrafast laser material is titanium-doped sapphire or Ti:sapphire, thanks to its large bandwidth and broad tuning range; turnkey commercial Ti:sapphire lasers can deliver pulses as short as 6-10 fs (10×10^{-15} s). Ti:sapphire lasers are universally pumped using a green-wavelength CW pump laser, usually a frequency-doubled OPSL or Nd-based laser, but the future availability of blue or green high-power diodes may shift this balance.

With cavity lengths of about 1.5 to 2 m, typical repetition rates are 75-100 MHz, and the average power can reach 4 to 5 W. The peak power of a mode-locked 4-W Ti:sapphire laser operating at 100 MHz and producing 100-fs pulses is then $4/(100 \times 10^{-15} \times 100 \times 10^6) = 400$ kW.

It is important to note that there is simply no other man-made source that can produce events on such a short timescale, a timescale that is faster than the formation and dissociation of chemical bonds in molecules. Moreover, the high peak power of these lasers enables the generation of many nonlinear processes; that is, processes where two, three or more photons simultaneously interact with a target. Thanks to the short pulse duration and high peak power, the advent of femtosecond lasers in the 1990s enabled a true revolution in science that resulted in new areas of research as diverse as femtochemistry or nonlinear optical microscopy. In addition to scientific applications, pico- and femtosecond lasers are now employed in industrial materials processing applications that require ablation or materials modification without any residual thermal effect and/or on a submicron spatial scale.

In the past few years, new laser materials have emerged that can produce femtosecond-class pulses. These lasers are mostly based on Yb atoms in various host crystals (YAG, KGW, KYW, etc.). One important distinction between Ti:sapphire and Yb lasers is that Yb lasers are limited to producing pulses of ~100-200 fs, while Ti:sapphire can achieve <10 fs; secondly, Ti:sapphire lasers are tunable over as much as 400 nm (680-1080 nm), while Yb lasers can be tuned only over 10-20 nm at the most (1040-1060 nm). This means that ultimate tuning range and short pulse duration reside with Ti:sapphire lasers. On the other hand, the simplicity of directly diode-pumped Yb lasers and their capability to produce higher average powers (in excess of 100 W) means that industrial applications will be best addressed by the laser type.

Yb also lends itself for use as an active medium for ultrafast fiber lasers. These lasers use a diode-pumped optical fiber with an integrated mode-locking mechanism to produce 100-fs to 10-ps pulses with wavelengths around 1 mm, while fiber lasers based on Er produce wavelengths around 1550 nm and are now a common staple in optical fibers for telecom applications.

Mode-locked lasers produce pulses at a high repetition rate and with average

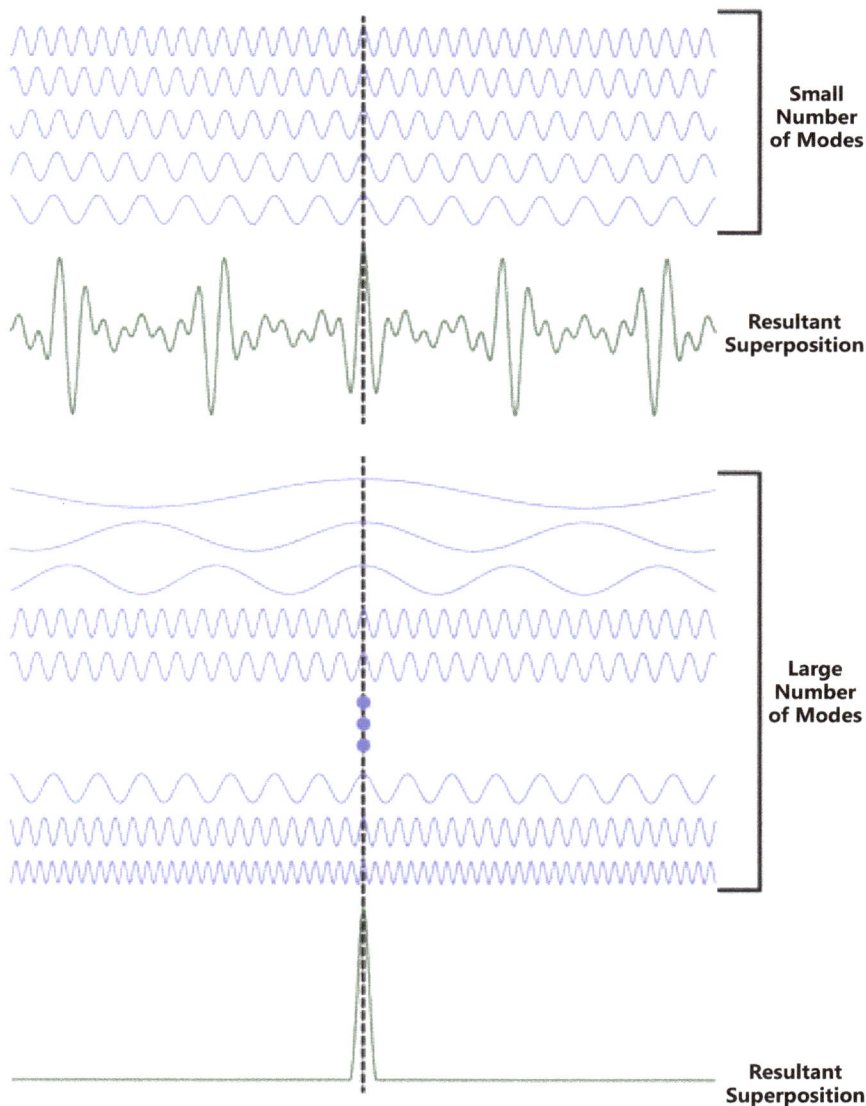

Figure 7. When a very large number of laser modes that all have a 'zero' in the same position interfere, the resultant superposition is an extremely narrow pulse.

powers of 1-100 W. The high repetition rate, necessary to support mode-locking, limits the energy per pulse to tens or hundreds of nanojoules. For some (mostly scientific) applications, it is desirable to have millijoule- or even joule-level energies with peak powers of hundreds of megawatts up to a petawatt. This can be achieved through chirped pulse amplification (CPA) (Figure 8). CPA is a technique that uses a mode-locked oscillator as a first building block. The output of this megahertz-repetition-rate oscillator (~20-100 fs) is stretched to become ~50-200 ps. A fraction of these 50- to 200-ps pulses is injected into an amplification stage, where the pulses are amplified to the millijoule level. At this point, the pulses are ejected from the amplification stage, usually with an EOM, and become recompressed close to the original pulse duration of 30-100 fs or less. The word "chirped" refers to the fact that the pulses are stretched to longer duration to avoid damage to the optical components of the amplifier stage caused by the high peak power. The amplifier of course requires optical pumping to sustain gain and amplification; in the case of Ti:sapphire, most CPA systems operate at 1-10 kHz or ~100 kHz. The lower-repetition-rate systems are energized by a Q-switched pump laser, while the higher-repetition-rate ones are pumped by CW green pump lasers, of the same type used to pump a mode-locked Ti:sapphire laser. Yb systems can also use a CPA scheme but, because of the diode pumping, they are more suitable to operation at higher repetition rates and lower energy per pulse than Ti:sapphire systems.

Another distinct class of mode-locked, ultrafast lasers has been developed for industrial uses. Scientific ultrafast lasers are typically designed to deliver cutting-edge performance in at least one of three parameters: namely, pulse duration, pulse energy or repetition rate. In contrast, industrial ultrafast lasers are usually designed to produce a balance between these characteristics that enables materials processing at practical rates, while also delivering high operational reliability at the lowest possible cost. Typically, these industrial ultrafast lasers have pulse widths in the 10-ps range and average powers that range from about 10 up to 100 W.

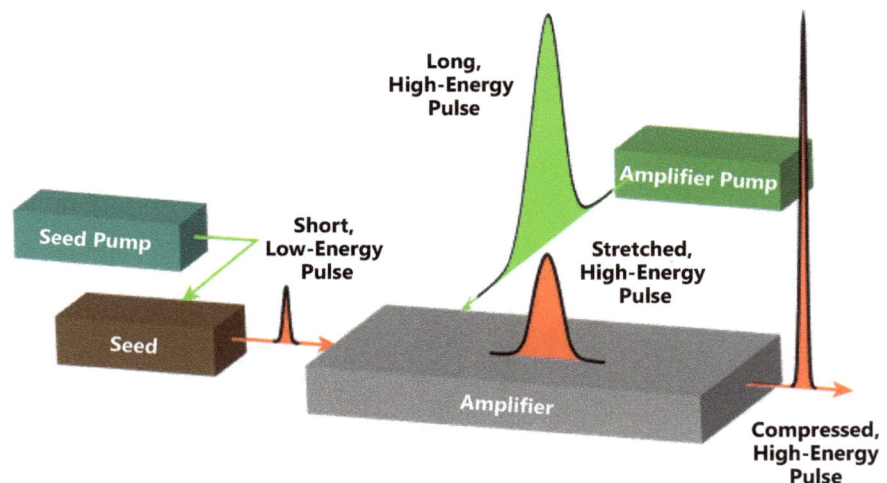

Figure 8. The basic elements and operation of a chirped pulse amplifier (CPA).

Frequency doubling and harmonic generation

Even with the broad choice of commercially available lasers, it is not always possible to find one that exactly matches the wavelength required by a specific application. Ti:sapphire lasers are broadly tunable, but in most cases, they are too complex for industrial applications and unable to reach the all-important UV region of the spectrum. OPSLs are simple and can be designed at many wavelengths in the 920- to 1160-nm region but are not ideal for pulsed operation. To achieve the desired wavelength in just about any regime of operation — CW, pulsed or ultrafast — the processes of harmonic frequency conversion and parametric generation provide wavelength flexibility when used in conjunction with the lasers described so far. All these processes are related and are called nonlinear phenomena since they depend nonlinearly on the laser peak power. That is, they are proportional to the square, third or higher power of the laser output power.

In simple terms, when an intense and/or tightly focused laser beam passes through a suitable crystal, its oscillating electric field interacts with the electrons of the crystal in several ways. One of these mechanisms distorts the electron cloud in the crystal, thereby polarizing the atoms at a frequency that is the same as that of the laser beam, but also at a frequency that is its double (nonlinear polarization). This frequency corresponds to a wavelength that is half that of the incoming laser. The nonlinear polarization is much smaller than the linear term, but it depends on the square of the laser power, therefore increasing more strongly in the presence of an intense laser pulse. It generates an optical field at double the frequency of the original laser beam, with the result that part of the incoming laser power will be converted to half the original wavelength (second-harmonic generation (SHG) or frequency doubling) (Figure 9). Since energy has to be conserved, any gain in the SHG beam is traded for a decrease in power of the original beam. In some cases, it is possible to achieve an almost total conversion of the original ("fundamental") beam into its second harmonic. Common crystals for SHG are BBO, LBO and KDP. The most common example of SHG is the conversion of a Nd-based laser IR output at 1064 nm into a green output at 532 nm (green), constituting the most popular visible wavelength, used ubiquitously to pump Ti:sapphire lasers.

Efficient SHG can be achieved only under a condition called "phase matching." Under most conditions, the light at the new frequency would be reconverted to the original frequency and lost or simply not added in phase to create any sizable power. This difficulty is overcome by choosing a crystal temperature and orientation that create the so-called phase-matching condition where the phase velocities of the fundamental and second-harmonic light are the same. This happens by choosing a specific direction of propagation (usually a function of temperature and wavelength) in the crystal such that the two waves propagate at the same velocity.

Extensions of the SHG process are third-harmonic generation (THG), where the wavelength at one-third the incoming wavelength is created by the interaction of an SHG beam with its fundamental; and fourth-harmonic generation (FHG), where the SHG beam is frequency-doubled again. All of these harmonic processes can be generalized as frequency-mixing, where two coherent beams at different wavelengths are mixed to produce sum-frequency and difference-frequency generation (SFG and DFG, respectively).

Harmonic generation can be applied to CW, pulsed and ultrafast lasers, greatly expanding the range of available wavelengths. Pulsed or ultrafast lasers have enough peak power (kilowatt range) to achieve relatively high conversion efficiency in a single pass through the harmonic crystal. On the other hand, CW lasers usually do not produce sufficient power for efficient harmonic generation, so that the power in the crystal has to be enhanced by putting the nonlinear crystal inside the laser cavity ("intracavity doubling"), or building a cavity ad hoc around the crystal ("resonant doubling") that matches the modes of the original CW laser cavity.

Optical parametric generation

In the SFM process described in the previous section, two photons at different frequencies interact to produce a single photon at a frequency that is the sum of the two initial frequencies. The opposite process is also possible: A single photon interacts with a suitable crystal and disappears, creating two photons of lower and different (called "nondegenerate") energies. This process, called optical parametric generation, is useful because it results in the generation of two new tunable wavelengths, bound solely by the conservation of energy and momentum, and by the nonlinear crystal refractive indices n:

$$\nu_p = \nu_s + \nu_i$$
$$1/(n_p \lambda_p) = 1/(n_s \lambda_s) + 1/(n_i \lambda_i)$$

The subscripts p, s and i refer to the pump wavelength and to the two new wavelengths that are called (for historical reasons) "signal" and "idler," with the signal wavelengths being the shorter of the two and both being longer than the pump wavelength. Phase matching takes place when the refractive indices of the three

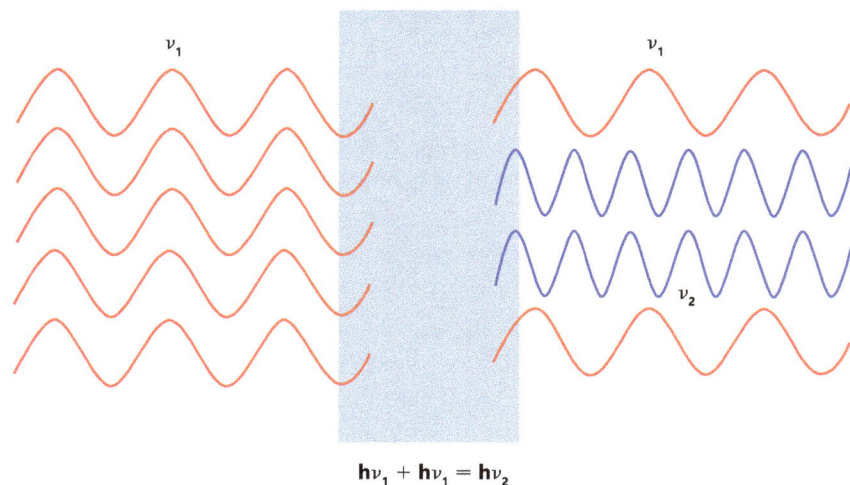

Figure 9. Basic functioning of a second-harmonic generating crystal.

$$\mathbf{h}\nu_1 + \mathbf{h}\nu_1 = \mathbf{h}\nu_2$$

wavelengths np, ns and np satisfy the above momentum conservation equation in the crystal. This can be accomplished by changing the temperature or angle of the crystal, or the pump wavelength (if the pump laser is tunable) so that phase matching takes place at the desired signal or idler wavelength.

While the wavelength of an SHG beam is unambiguously determined by the pump wavelength, the OPG process can generate an infinite set of wavelength pairs. Amplification of the desired wavelength pair requires not only phase matching in the crystal but also "jump starting" the process from the noise of the statistical distribution of wavelength pairs. This is exactly what happens in an optical parametric oscillator (OPO) or an optical parametric amplifier (OPA), both advanced laser accessories able to produce tunable outputs anywhere from the mid-UV to the mid-IR.

In an OPO, the signal and possibly the pump are resonated in a laser-like cavity, where the desired signal wavelength starts from the noise (random distribution of signal/idler pairs) and is amplified by passing through the crystal properly phase-matched for that wavelength at the same time as the pump during each round-trip (Figure 10). In an OPA, a sapphire or YAG disk is pumped to produce a relatively bright beam of white light that, as the name suggests, contains all the wavelengths of the visible and near-IR spectrum. The unique wavelength pair that is phase-matched in the OPA crystal will then be amplified when the crystal is pumped by the pump laser.

Nano-, pico- and femtosecond OPOs are complex devices that are implemented in conjunction with pulsed and ultrafast pump lasers. CW OPOs are equally, if not more, complex. OPAs are easier to design and build but require a more energetic pump to produce the white light and one-pass amplification in the crystal. For this reason, they are pumped by CPA pico- or femtosecond amplifiers producing at least several microjoules. The addition to an OPA/OPO of one or more stages of harmonic generations and mixing yields a range of wavelengths that can cover 200 nm to 20 μm.

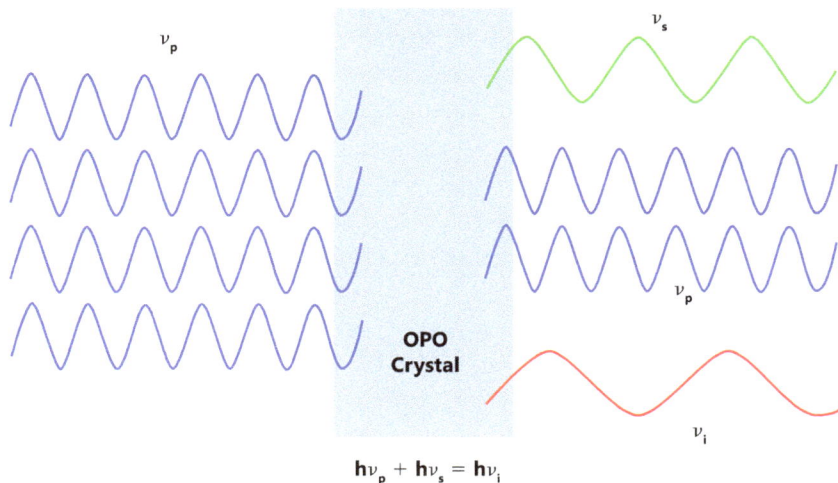

$$h\nu_p + h\nu_s = h\nu_i$$

Figure 10. An optical parametric oscillator (OPO) converts an input photon into two photons having lower energy, and which conserve the energy and momentum of the input.

Common laser types

For many years, the most common CW laser was the helium neon laser, or HeNe. These low-power lasers (a few milliwatts) use an electric discharge to create a low-pressure plasma in a glass tube; nearly all emit in the red at 633 nm. In recent years, the majority of HeNe applications have switched to visible laser diodes. Typical applications include barcode readers, alignment tasks in the construction and lumber industries, and a host of sighting and pointing applications ranging from medical surgery to high-energy physics.

In fact, the laser diode has become by far the most common laser type, with truly massive use throughout telecommunications and data storage (e.g., DVDs, CDs). In a laser diode, current flow creates charge carriers (electrons and holes) in a p-n junction. These combine and emit light through stimulated emission. Laser diodes are available as single emitters with powers up to tens of watts, and as monolithic linear bars with numerous individual emitters. These bars can be assembled into 2D arrays with total output powers in the kilowatt range. They are used in both CW and pulsed operation for so-called direct diode applications. But even more importantly, laser diodes now underpin many other types of lasers, where they are used as optical pumps that perform the initial electrical-to-optical power conversion.

For example, higher-power visible and UV CW applications were originally supported by argon-ion and krypton-ion lasers. Based on a plasma discharge tube operating at high current, these gas-phase lasers are large and inefficient, generating a large amount of heat that must be actively dissipated. The tube also has a finite lifetime and thus represents a costly consumable. In most former applications, the ion laser emitting at blue or green wavelengths was displaced by DPSS lasers. Here, the gain medium is a neodymium-doped crystal (usually Nd:YAG or Nd:YVO$_4$) pumped by one or more laser diodes. The near-IR fundamental at 1064 nm is then converted to green 532-nm output with the use of an intracavity doubling crystal.

The DPSS laser, in turn, has been challenged by several newer technologies, with the OPSL the most successful of these. Here the gain medium is a large-area semiconductor laser that is pumped by one or more laser diodes. The OPSL offers numerous advantages, most notably wavelength and power scalability. Specifically, these lasers can be designed to operate at virtually any visible wavelength, at last freeing applications from the restrictions of limited legacy-wavelength choices (i.e., 488 and 514 nm from argon-ion lasers and 532 nm from frequency-doubled YAG lasers). Indeed, OPSLs represent a paradigm shift in lasers because they can be designed for the needs of the application instead of vice versa.

OPSL is now a leading technology in low-power bioinstrumentation applications, most notably at 488 nm; the power scalability and inherent low noise of OPSL technology is now seeing multiwatt green and yellow OPSLs moving strongly into other applications, including scientific research, forensics, ophthalmology and light shows.

While YAG and other neodymium crystal hosts lend themselves to operation in

CW, Q-switched and mode-locked operations, laser diodes, OPSL and ion lasers do not support Q-switched operation and are virtually not used in mode-locked regime.

At longer wavelengths, carbon dioxide (CO_2) lasers, which use plasma discharge technology, emit in the mid-infrared around 10 µm. Most are CW or pseudo-CW, with commercial output powers from a few watts to several kilowatts. A similar technology is the carbon monoxide (CO) laser, which was originally developed in the 1960s, but only made truly practical for industrial use in 2015. CO lasers emit in the 5- to 6-µm spectral range. This shorter wavelength, mid-infrared output offers two important advantages for some applications as compared to CO_2 lasers. The first is that many metals, films, polymers, PCB dielectrics, ceramics and composites exhibit significantly different absorption at the shorter wavelength, which can sometimes be exploited to advantage. The second is that they can be focused to smaller spot sizes due to diffraction, which scales linearly with wavelength. Together, these characteristics enable the CO laser to deliver superior results in some glass processing, film cutting and ceramic scribing applications.

Another important technology is the fiber laser, which can be operated in CW, Q-switched and mode-locked formats. Here, laser diodes optically pump a rare-earth doped fiber, which typically emits at about 1 µm.

Nd:YAG, CO_2, fiber and direct diode lasers are the workhorses of industrial laser applications. Direct diode lasers predominantly service low-brightness applications, such as heat treating, cladding and some welding applications. This is because direct diode lasers offer the lowest capital cost of any industrial laser type, as well as the lowest operating costs, due to their high electrical efficiency. On the downside, high-power laser diodes or arrays cannot deliver anything close to the diffraction-limited beam provided by other laser types.

The advent of slab-discharge technology has allowed the size:power ratio of CO_2 lasers to be greatly scaled down, increasing their utility in subkilowatt applications. Low-cost waveguide designs also support a healthy market for CO_2 lasers with powers in the tens of watts, primarily in marking and engraving applications.

Sealed CO_2 lasers and fiber lasers have come to dominate metal cutting in the 2- to 4-mm thickness range. Sealed CO_2 is usually the first choice when both metals and nonmetals must be processed, while fiber lasers have proved quite successful in certain markets that can benefit from their combination of high repetition rate, low pulse energy and high brightness. They also excel at metal cutting and welding in the 4- to 6-mm thickness range, as well as some marking applications. Flowing gas CO_2 lasers still dominate the market for cutting thick metal (>6 mm).

Nd:YAG can deliver the high peak power for materials processing applications such as metal welding. In these heavy industrial applications, raw power is more important than beam quality, and for many years, these lasers were lamp-pumped. But the ever-increasing power and lifetime characteristics of laser diodes are causing these lasers to switch to diode pumping; i.e., DPSS lasers.

Conversely, lower-power Q-switched DPSS lasers are often based on Nd:YVO$_4$. These are usually optimized for high beam quality for micromachining and microstructuring applications with high repetition rates (up to 250 kHz) to support high throughput processes. They are available with powers up to tens of watts with a choice of near-IR (1064 nm), green (532 nm) or UV (355 nm) output. The UV is popular for producing small features in "delicate" materials because it can be focused to a small spot and minimizes peripheral thermal damage. Deep-UV (266 nm) versions are starting to be used in some applications, but their relatively high cost and the need for specialty beam delivery optics causes many potential applications to rely instead on 355-nm lasers optimized for short pulse duration, which can produce similar results in many materials.

Excimers represent another important pulsed laser technology. They can produce several discrete wavelengths throughout the UV; depending on the gas combination, emission ranges from 157 to 348 nm. The deep-UV line at 193 nm is the most widely used source for lithography processes in the semiconductor industry. The 308-nm wavelength is used for annealing silicon in high-performance displays. The same wavelength is also key to generating a unique long-wear surface on the cylinder liners of high-performance diesel engines. And finally, excimers have a unique ability to produce high pulse energies — up to one joule per pulse. This enables direct writing of low-cost electronic circuits for applications such as medical disposables.

Ultrafast lasers for scientific applications are dominated by Ti:sapphire, as already described. Ultrafast lasers are also a fast-growing technology for micromachining and other high-precision materials processing applications. While there is some diversity in the form and construction of commercially available industrial ultrafast lasers, they all utilize a certain basic configuration. Specifically, a passively mode-locked oscillator is used to generate output at the pulse widths of about 10 ps or shorter that are necessary to drive photoablation. However, most mode-locked oscillators produce relatively low energy pulses (in the nanojoule range) at repetition rates in the tens of megahertz. Best results for micromachining are achieved when the pulse-to-pulse overlap is in the range of 50 to 70 percent. In other words, the beam deflection mechanism moves the beam about one-third of the beam diameter before the arrival of the next ultrafast pulse. Consequently, a repetition rate in the range of tens of megahertz is too high to be used with existing scanning technology, so a pulse picker selects a fraction of these pulses. The energy of these pulses is then boosted in an amplifier to produce the final output.

Most commercial picosecond products are based on one of the following architectures:
- A fiber laser oscillator followed by a fiber- or rod-type amplifier
- A fiber laser oscillator followed by a free-space amplifier
- A diode-pumped solid-state oscillator followed by a free-space amplifier

The all-fiber (oscillator and amplifier) approach has the advantage of being relatively low cost and holds the potential of being robust. The big negative is that non-linearities, scattering and other effects in the fiber amplifier limit the maximum per-pulse energy that can be attained to about 10 μJ (at a 10-ps pulse width). This level of pulse energy can cater to some applications, but a large number of applications are served with pulse energies in the 100-µJ range. One can use specialty fibers to increase the mode inside the fiber and thereby allow for larger pulse energies, but such fibers lead to limited bend radii and hence bring their own packaging limitations.

To achieve the higher pulse energies required for most applications, a fiber oscillator can be mated with a free-space amplifier. Because of the relatively low energy output from the fiber seed, a regenerative amplifier is often used. In a regenerative amplifier, a pulse undergoes a large number of passes and can therefore be amplified substantially. Regenerative amplifiers also have the advantage of being compact and delivering good beam performance.

The third approach is to use a diode-pumped solid-state oscillator (usually with Nd:YVO$_4$ as the gain medium), which can produce higher pulse energies than a fiber seed. This is followed by a free-space amplifier, typically in either a regenerative or multipass configuration. In fact, more than one amplifier stage can boost power to levels as high as 100 W.

And finally, many other types of niche and exotic lasers exist that are beyond the coverage of this overview article. Examples include Raman lasers used in telecommunications, quantum cascade lasers used in some gas-sensing applications, and chemical lasers, which tend to be limited to military programs.

Historic Lasers Find New Life in Niche Applications

With all the hoopla surrounding new high-brightness lasers such as disk and fiber, some think the death knell is sounding for CO_2, lamp-pumped and other older, well-established types. However, the lack of a one-size-fits-all laser means these classics are far from giving up the ghost.

BY DAVID HAVRILLA, TRUMPF INC.

In the years since the introduction of high-power disk and fiber lasers, we have witnessed not only the resilience of the historic industrial lasers, but also enormous advances in direct-diode lasers. In fact, the direct-diode laser, which preceded disk and fiber, is certain to be somewhat disruptive to the current high-brightness leaders in the not-too-distant future. So why have the historic lasers survived the cut? And why should disk and fiber not rest on their laurels?

The real key to survival is that manufacturers and users alike are realizing that there is no such thing as a universal laser — there is no Swiss Army knife of lasers. And if you ponder it, even the Swiss Army knife isn't as universal as we might think — try using it to drive a nail! In any case, there seem to be niche applications in which each laser type can flourish.

Stainless steel tube welding with a CO_2 laser. All images courtesy of Trumpf Inc.

Housing – spot-welded in burst mode.

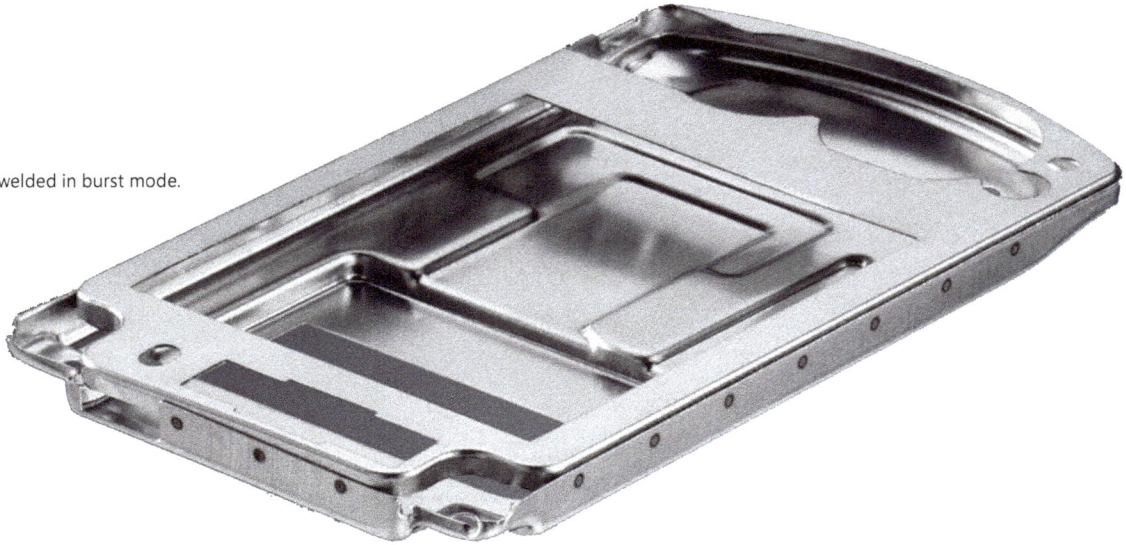

Pulsed lasers

With the advent of low-power fiber lasers, what can pulsed lasers offer? Plenty, it turns out. For one thing, by their very nature, they can emit extremely high peak powers. For example, a 100-W average power pulsed laser can deliver a whopping 8 kW or more in processing power to the workpiece. So, for the price of a 100-W laser, the user benefits from 8000 W of processing power. This high peak power is useful for penetrating deep into metal and is especially beneficial for highly reflective materials such as aluminum.

Oftentimes, especially when welding, a relatively large focused spot is needed to ensure sufficient fusion or to accommodate inadequate part fit-up. Fiber lasers are extremely focusable; that is, they focus to a very small spot. Although the spot can be enlarged through special optics or by launching into a large process fiber, the lack of peak power limits their usefulness. Quasi-CW fiber lasers can address the lack of peak power, but at the cost of more diodes, because fiber lasers do not exhibit a higher-than-CW power when pulsed. In other words, if an 8-kW pulse is needed, 8 kW of fiber laser power is required.

Another area where pulsed lasers shine is spot welding. There are pulsed lasers (including Trumpf's TruPulse series) that are equipped with a special operating feature called burst mode. In burst mode, all of the stored energy in the laser power supply can be used to lay down myriad spot welds in rapid-fire succession. When the joining is done, and while the welded parts are being exchanged, the laser is recharging back to full capacity, getting ready for the next burst of welds. This may not sound very impressive, but it is, and here's why: With burst mode, the laser can put out a higher average power during the "bursts" than what the laser is rated for — in anticipation of the rest it will get during recharge. The user can therefore save money by purchasing a laser with lower average power than what would normally be required.

So for applications requiring high peak power, welding with a larger focus or rapid-fire spot welding, pulsed lasers are still the most cost-effective and ideally suitable on the market. More could be said, but let's move on to the longtime staple of the laser industry.

CO$_2$ lasers

Not merely tried and proved thousands of times over, the CO$_2$ laser is the oldest industrial laser with the largest installed base of any laser type. It continues to reign over several applications. We can start with the obvious niches: plastics, glass and fabric. This is all to do with wavelength. The longer wavelength of the carbon dioxide laser (10.6 μm) simply absorbs much better into these materials compared with the disk or fiber lasers. Of course, the 1-μm lasers have their absorption niches as well, and they open the door to processing highly reflective materials such as copper and brass.

But let's consider the most common industrial materials: steel, stainless and aluminum. When cutting thin sheet metal, the 1-μm wavelength produced by the disk and fiber lasers does have a speed advantage. The reason is somewhat complicated and debated, but it's fair to say that it lies somewhere in the realm of wavelength and keyhole geometry — i.e., incidence angle — because of their influence on absorption. For oxygen cutting of mild steel, the 1- and 10.6-μm cutting results are comparable over about 1.5 mm. However, for nitrogen cutting of stainless with thicknesses greater than 4 mm, speeds are similar, but the cut-quality scepter is passed to the CO$_2$ laser. Not only is the roughness worse, but another detrimental thing happens when performing oxide-free cutting of thick stainless with 1-μm lasers: burr. Not the frigid, "I'm really cold" kind of burr, but the "ragged edge at

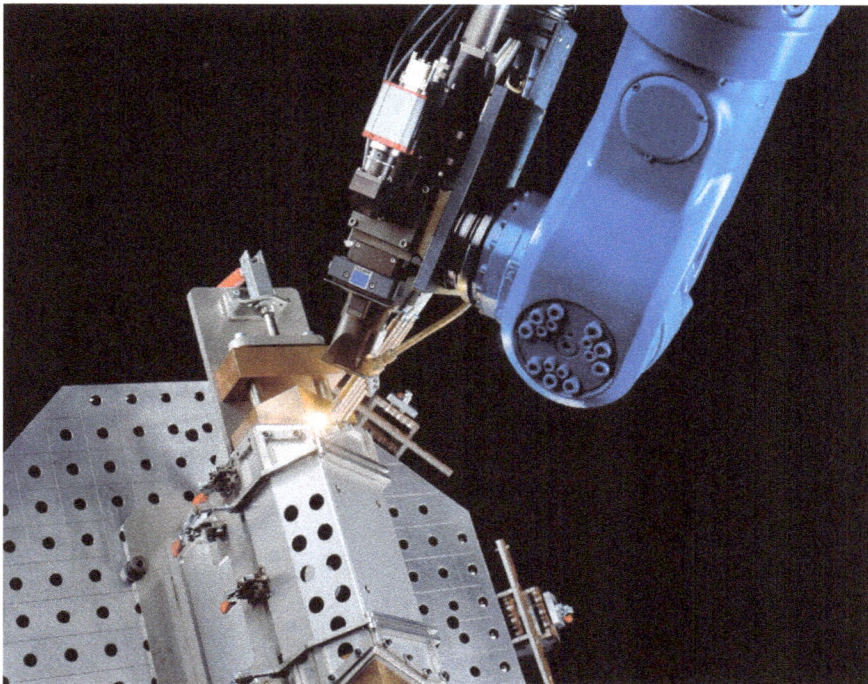

Laser welding of an enclosure housing with a fiber-delivered direct-diode laser.

the bottom of the cut that's difficult to remove" kind of burr. Not only is there a tenacious burr, but the roughness of the cut is worse than that produced with the CO_2 as well. When cutting aluminum, 1 µm nets the higher speed, but the CO_2 laser offers less burr at thicknesses greater than 3 mm.

What about welding? The carbon dioxide laser has an advantage here as well. When welding with high-power disk and fiber lasers, more expulsion (spatter) is generated compared with the carbon dioxide laser, probably tied to the absorption influences mentioned above. The difference in the amount of spatter can be significant or insignificant, depending upon the product that is being welded, and the weld depth and speed. In general, the deeper the weld penetration, the greater the amount of weld spatter. One application where minimizing the spatter is crucial is laser welding of stainless steel tubes. These tubes, whether used in the food industry or for automotive exhaust pipes, cannot have spatter on the inside of the tube resulting from function, nor on the outside of the tube because of form. Here the CO_2 laser excels.

Another spatter-critical application is welding of power train components. Whenever you combine really hard small particles with high-speed rotating gears, you have a problem. Again, the CO_2 laser has the advantage. Please don't misunderstand. There are numerous welding applications — and some power train applications — where spatter generation is inconsequential, and there are ways to protect the component being welded — and the clamping device, focus optics, etc. — from the spatter. But the gold medal goes to the carbon dioxide laser for applications where spatter must be minimized.

Direct-diode laser

The direct-diode laser has been around since the 1990s. Its use has been somewhat limited because of its focusability — or lack thereof. You might think that the highly focusable, high-power disk and fiber lasers would send these lasers into obsolescence. Ah, but they are not quite ready to fade away. In fact, it's likely that the opposite will be true in the not-too-distant future. In the many years that direct-diode lasers have been around, there has been much progress in both the available power and focusability of these lasers. And for good reason — the diode is the heart of the disk and fiber lasers.

So how is it that the newcomers (disk and fiber) could one day jettison their glory, while the diode laser is poised for grandeur? The primary answer can be communicated with just three letters — WPE (wall plug efficiency). You've probably heard that the WPE of the disk and fiber laser is pretty good, around 25 percent, with all things considered. But the WPE of the direct diode approaches 40 percent and, in theory, even higher. Why is the direct diode so much more efficient than the glitzy newcomers? With the disk and fiber lasers, diodes are used to pump a lasing medium to create a laser beam. If we cut out the middleman (the lasing medium) and use the diodes directly, WPE skyrockets.

So where are direct diodes on the "excellent focusability at higher power" road map? It wasn't long ago when direct-diode lasers could be used only for brazing —

and even now you can't beat them for brazing. Today, direct-diode lasers are used for deep-penetration keyhole welding, and even cutting. The race is on to see who will be the first to deliver high-power diode lasers capable of remote welding; i.e., up to 6 kW delivered through a 200-µm fiber. All indications are that we will see that product within this decade — perhaps even within the first half.

Now the punch line: When the high-power, remote-welding-capable direct-diode laser makes it to the field with higher WPE, smaller footprint and lower cost, the disk and fiber had better start looking for their niches — and they will find them.

Conclusion

For "Star Trek" fans, the new high-brightness lasers are not the Borg. They will not assimilate all their foes; nor is resistance futile. In reality, there is no one type of laser that can accomplish all tasks equally well. There is strength in diversity. And even the lasers that now seem as though they could be the Borg — well, one day they, too, will settle into those applications where they excel.

Meet the author

David Havrilla is manager of products and applications at Trumpf Inc. in Plymouth, Mich.

Fiber vs. Disk: Which Laser Will Make the Cut?

Some in the industry say they see fiber continuing to make large gains in the manufacturing space, but others say don't count thin-disk out.

BY MELINDA ROSE, SENIOR EDITOR

The boom in fiber lasers that began several years ago has led some to question whether there is still room in the industrial manufacturing space for both the glamorous new star and its slightly less hip cousin, the disk laser. A short survey of companies on both sides — and neither side — of the issue reveals that the answer is yes, but that the choice is not limited to just one or the other.

To understand where these lasers are heading in manufacturing, we first have to look at where they've been.

The skinny on fiber

Fiber lasers are much older than disk lasers, having been conceptualized just months after the first laser — the ruby — was demonstrated by Ted Maiman in 1960. Working for American Optical, Elias Snitzer theorized the fiber laser in 1961 and demonstrated the first lanthanide-doped glass fiber laser in 1963. But that flash-

With laser outputs of up to 16 kW and beam qualities starting at 2 mm × 3 mrad, disk lasers offer unrivaled robustness — for highly productive thin sheet cutting, as well as for welding. Courtesy of Trumpf Group.

lamp-pumped laser had an extremely low output power, a condition that was to plague lasers for decades.

In fiber lasers, a rare-earth element such as erbium or ytterbium is doped into the core of an optical fiber. Laser emission is created within the fibers using a semiconductor diode as the light source and delivered through a flexible optical fiber cable. Fiber lasers have a monolithic, entirely solid-state design that does not require mirrors or optics.

Two important, interrelated developments have helped move fiber lasers from lab curiosity to serious commercial contender in the past decade: The creation of high-power laser diodes that boost the lasers' pump power enough to compete with other lasers, particularly less efficient lamp-pumped systems, and market advancements that have made those diodes more affordable and reliable.

Key advantages of fiber lasers include high (near-diffraction-limited) beam quality, high power generation, high reliability, low jitter and amplitude noise, turnkey operation and small size.

As fiber lasers have continued to grow in output power, they have increasingly become used in materials processing applications such as marking, printing, welding and cutting, as well as for micromachining, drilling, soldering and annealing, among others.

The company most credited with making fiber lasers commercially viable is Oxford, Mass.-based IPG Photonics, founded in 1990. IPG has aggressively and successfully been replacing older, less efficient lasers with fiber alternatives for high-power industrial applications. In December 2009, the company introduced four quasi-CW (continuous wave) fiber lasers with peak pulse powers of 750, 1500, 3000 and 5000 W designed to replace the older flashlamp-pumped, long-pulse YAG lasers that then had a large share of the world laser market. IPG's

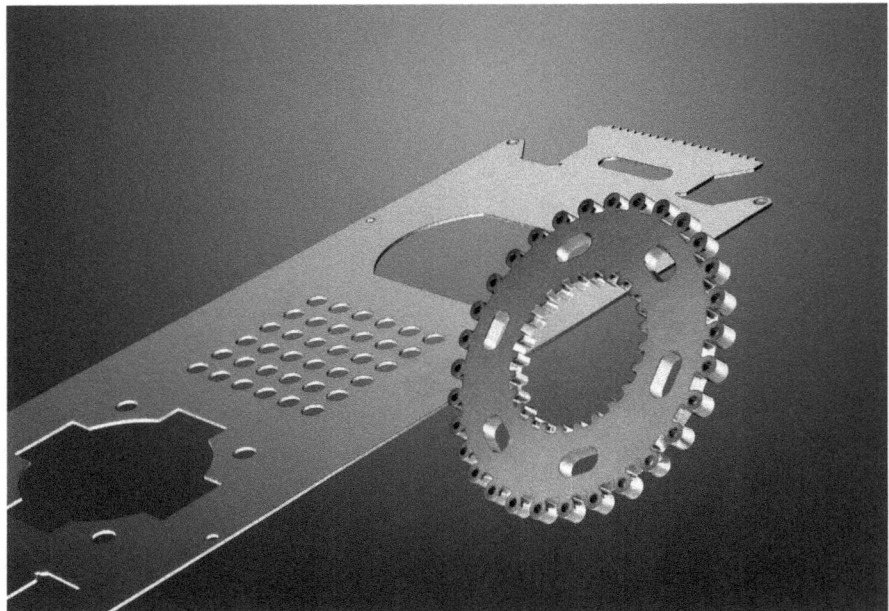

A laser beam enables one to cut nearly any shape, regardless of its complexity. Its strength lies in the ability to process the widest range of material types and thicknesses, distortion-free. Even foil-coated sheets are no match for a laser cutter. This flexibility is highly advantageous, especially when paired with a variety of machine models. The edges are smooth, and the processed parts are ready for assembly without requiring refinishing. Courtesy of Trumpf Group.

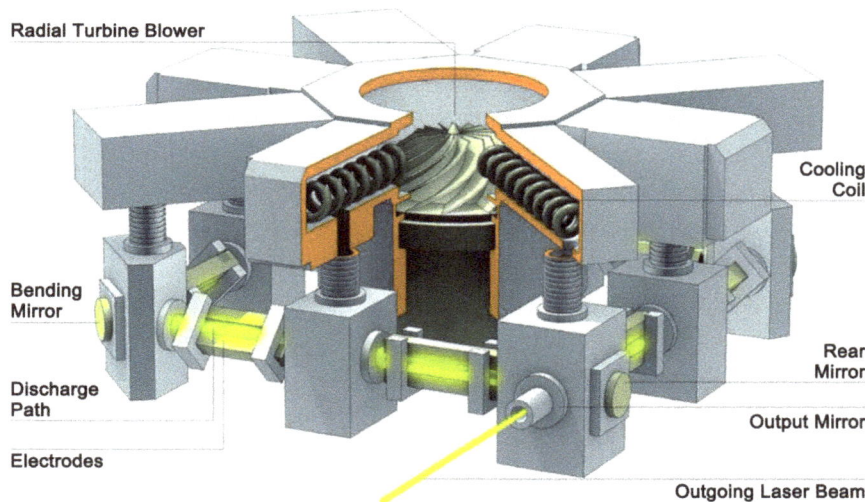

Radial Turbine Blower

Cooling Coil

Bending Mirror

Discharge Path

Electrodes

Rear Mirror

Output Mirror

Outgoing Laser Beam

Compact, powerful and reliable: a glimpse inside a flowing gas CO_2 laser with square design. Courtesy of Trumpf Group.

lasers offered efficiency rates of 30 percent compared with 3 percent for flashlamp-pumped lasers.

"We're replacing an awful lot of old lasers, but disk is trying to do the same thing," said Bill Shiner, IPG's vice president of industrial markets.

New kid on the block

Thin-disk lasers are diode-pumped, solid-state lasers first demonstrated in 1993 by Adolf Giesen and his group at the University of Stuttgart in Germany. The gain medium is a laser crystal, typically Yb:YAG, formed not as a rod but as a very thin disk, making the design less susceptible to distortion and other adverse effects as compared with many other laser types. Nd:YAG is also used but has a shorter emission wavelength. Other ytterbium-doped gain media are used for broad wavelength tuning. The thickness of the disk is usually much smaller than the laser beam's diameter.

In addition to its ability to be cooled very efficiently, the main advantages of thin-disk lasers are that their power and pulse energy can be scaled to much higher values than rods, fibers or slabs. It's easy to scale power with a thin-disk laser: You simply have to increase the diameter of the pump area of the disk, but the trade-off is decreasing beam quality.

Thin-disk lasers can generate ultrashort pulses at very high power levels, making them attractive for industrial applications. They can also generate high-energy nanosecond pulses with high beam quality, which is also attractive for some laser materials processing applications. Combining disks makes it possible for the lasers to achieve nearly any power level desired.

Continuing improvements to disk lasers are focused on increasing power and beam quality. "The target is getting more power out of one disk," said Dr. Kurt Mann, director of international sales at Trumpf, a major manufacturer of disk lasers. "We now have 6 kW of power available in one disk."

As recently as 2005, it took four separate disks to achieve an output of 4 kW, so improvements Trumpf has made to its thin-disk technology are also helping to make the lasers more affordable.

Team of rivals

The rivalry between fiber and disk lasers has been brewing for several years, as fiber lasers grow in power and efficiency, decrease in cost and move into manufacturing areas dominated by disk lasers, such as welding and cutting. Both types of lasers had marked advantages over, say, gas and crystal lasers in terms of scaling power not by increasing the gain medium but by increasing the power per volume.

Fiber lasers provide precise, fine welding and fast cutting, but disk lasers with their ease of power scalability can offer manufacturers flexibility. Most manufacturers seem to agree that the laser you choose is ultimately application-driven.

In the CW regime, arc-lamp-pumped lasers have been completely replaced with diode-pumped, Mann said, but even lamp-pumped lasers continue to have financial advantages for manufacturers.

At a Market Focus session on industrial lasers during CLEO (the Conference on Lasers and Electro-Optics) in May 2010, a brave soul posed the question, "What's better, thin-disk or fiber lasers?" A year later, the answer you get is pretty much the same, but it depends on who is doing the answering.

"If I look at fiber lasers, our price keeps going down and down. We're the least expensive multikilowatt laser on the market," Shiner said. "Automotive has completely gone to us worldwide."

"Both lasers have advantages in industrial applications," Mann said. "Fiber has excellent beam quality. When you have high-power applications and don't have high requirements on beam quality, then you go with the disk laser." Tasks such as car body welding and soldering, for example, don't require the high beam quality of a fiber laser.

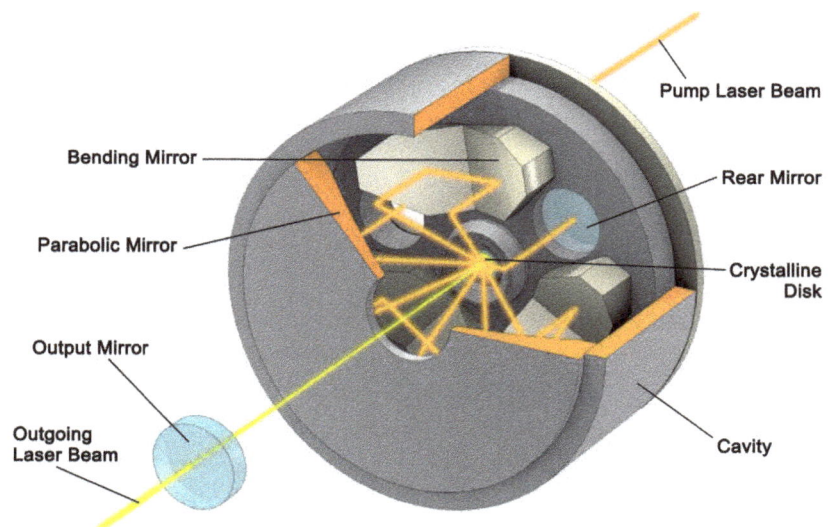

Transforming a low-quality pump laser beam into a high-quality beam: a glimpse inside a disk laser. Courtesy of Trumpf Group.

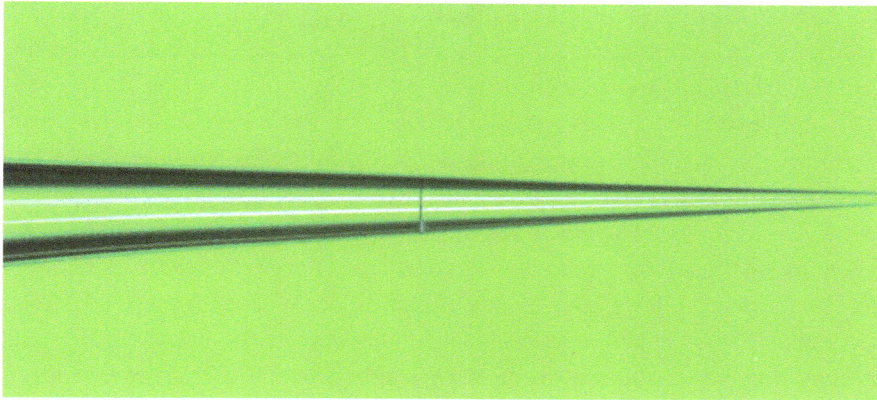

A 10-µm hole was machined into the side of this 120-µm glass needle used in life sciences applications. Courtesy of Raydiance.

"There is not one single laser which will cover all regimes," said Dr. Hans-Dieter Hoffmann of Fraunhofer Institute for Laser Technology (Fraunhofer ILT) in Aachen, Germany, adding that the bottom line with manufacturers will always be "what is the most cost-efficient source to do the process."

"The electro-optical efficiency of both lasers is very comparable, about 30 to 35 percent," Hoffmann said. "The real domain of fiber is in CW from zero to one kilowatt."

At Laser World of Photonics 2011 in Munich, Coherent — a major maker of CO_2 (carbon dioxide) and diode lasers — included in its floor models its first fiber laser, the 1-kW Highlight 1000FL, which is scheduled to officially launch this fall.

"We started with technology and ideas and then looked into the market and found out what the market wants," said Fred Kallweit, head of medical laser sales for Jenoptik Laser GmbH's Lasers and Material Processing Div. Jenoptik, of Jena, Germany, sells both thin-disk and fiber lasers. While many of its customers are

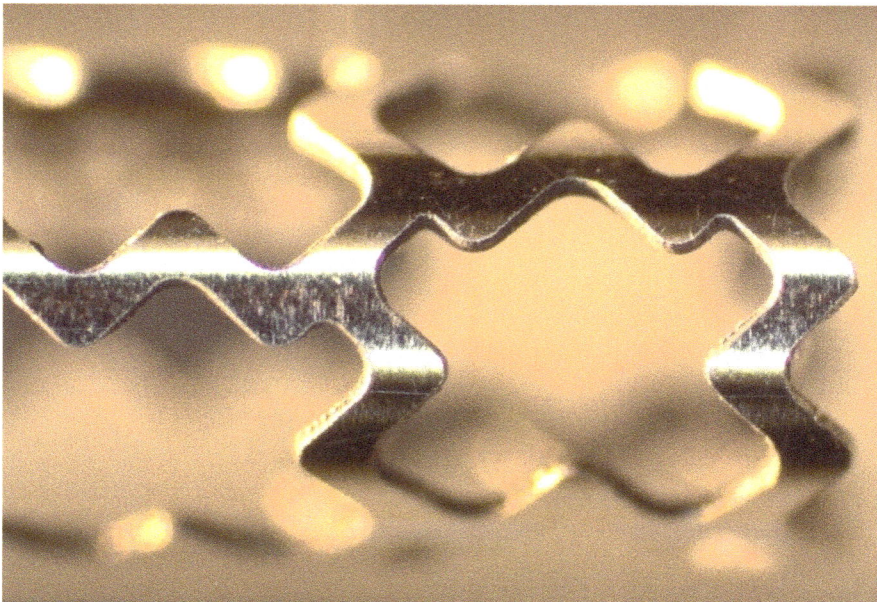

This Nitinol stent, machined with the StarCut Tube Femto, was cleaned in an ultrasonic bath of isopropyl alcohol for 5 minutes, then photographed. Note that there are no heat-affected zones. Courtesy of Business Wire.

interested in fiber, "We think the market is there also for thin disk," he said.

Blurring the line between the two laser types is the fiber-guided disk laser; the TruFiber system by Trumpf is one example.

Trumpf announced at Laser Munich 2011 that it had received an order worth €37.5 million from Volkswagen AG, mostly for 50 of its TruLaser Cell 8030 fiber-guided disk laser machines, to be delivered over the next two years. Volkswagen will use the machines in high-volume production of hot-formed parts for car bodies.

Also at the event, Trumpf displayed its new 2in1 fiber developed for its TruDisk disk lasers, which allows users to switch applications without changing the laser light cable. The cable has an inner-diameter fiber core of 100 µm for cutting and an outer ring with a 400-µm-diameter fiber for welding. The cable is a standard feature on the company's TruLaser Cell 3010 machine for 2D and 3D laser processing.

No KO to CO_2

A carbon dioxide laser electrically stimulates a gas-filled tube (a mix of helium, nitrogen and carbon dioxide) to produce light. It currently has no rivals in the industrial market for flatbed cutting of materials thicker than 4 mm, and the laser has replaced conventional methods such as punching or milling for materials with thicknesses up to 25 mm. It is also the most popular choice for welding thick plates.

Low-power CO_2 lasers are cost-efficient for marking, engraving and cutting applications where long wavelength is a plus. Although solid-state lasers have been making inroads, gas lasers continue to hold on to about half of the materials processing market share.

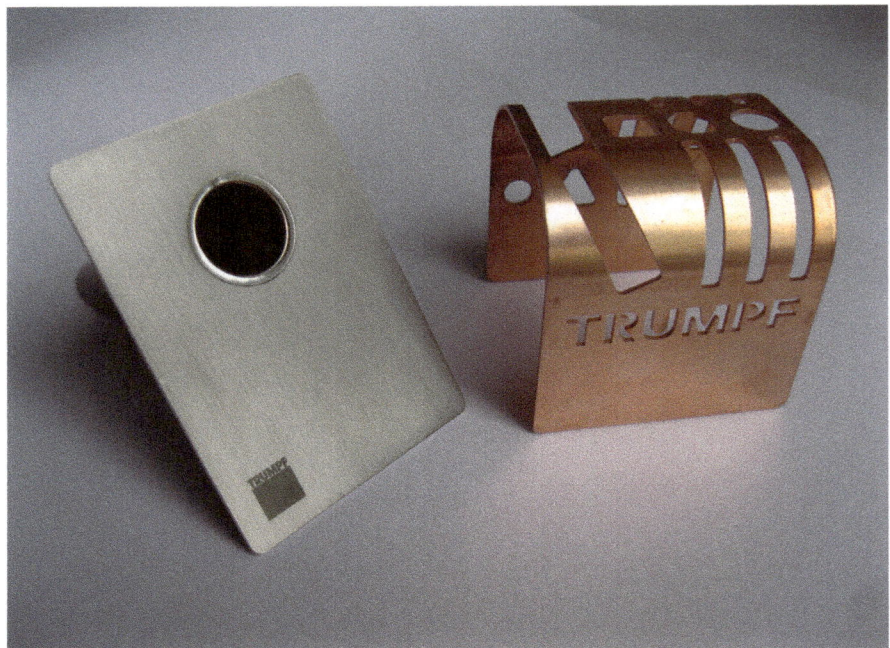

The five-axis Trulaser Cell 3010 is typically used for welding and cutting sample parts and prototypes. Courtesy of Trumpf Group.

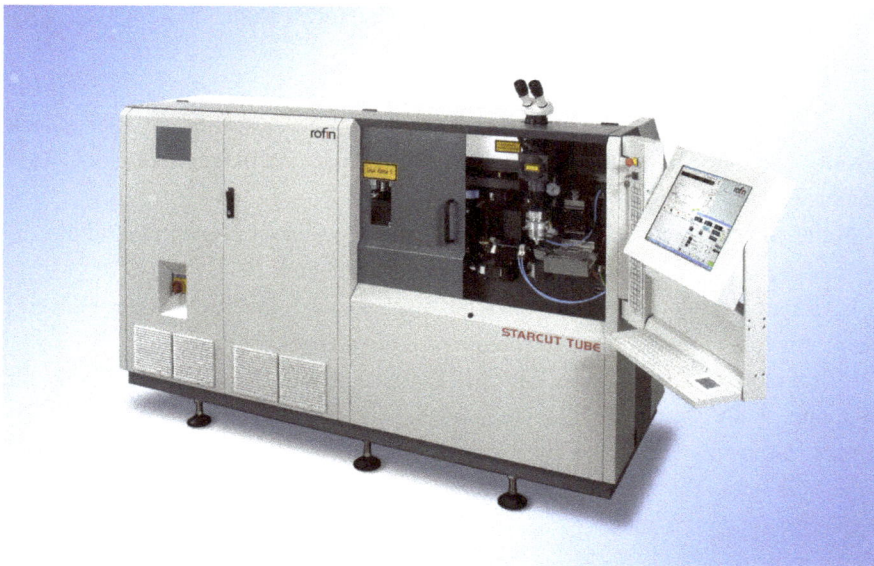

The new StarCut Tube Femto combines Rofin's systems solution engineering with Raydiances's femtosecond laser platform. This first rollout of Rofin systems with photons by Raydiance is designed for the cardiovascular stent manufacturing market. Courtesy of Business Wire.

Because the laser radiates in the far-infrared at 10.6 µm and most materials absorb at that wavelength, there is interest in using CO_2 lasers to cut materials besides metal.

"CO_2 lasers have a benefit in that they are a proven technology," Hoffmann said. "Because CO_2 is 10 µm, you can do plastic cutting with it. Glass is black for CO_2 emission and also at the component level [the lasers are used to drill holes in printed circuit boards]," Hoffmann said.

Both fiber and disk lasers emit at the 1-µm wavelength and currently can't compete with the cut quality of CO_2 lasers when it comes to thick materials, although the difference that wavelength makes in materials processing isn't fully understood.

"We're doing fundamental analysis on why 1-µm cutting is different from 10-µm cutting, regardless of the type of laser," Hoffmann said. For example, 10-µm lasers are twice as fast at cutting into components as 1-µm lasers.

"There are interesting applications for CO_2; I don't think CO_2 is at an end," he said.

At Laser Munich, Coherent presented the Metabeam 1000 machining tool, which includes a 1000-W, completely sealed CO_2 laser that it said competes favorably with fiber on a wide range of metals but can also process organic materials such as wood and plastic. The company also plans to release a kilowatt fiber-based Metabeam later this year.

An ultrafast future

As components and other devices continue to shrink in size, and as new and exotic materials are introduced into manufacturing, opportunities continue to open up for ultrafast laser applications in the life sciences, energy and advanced materials fields. Femtosecond and picosecond pulse repetition rates allow for athermal ablation — the removal of material without incurring heat that could induce melting or form burrs or cracks in the material.

"Fiber ultrafast lasers are starting to get introduced in large application areas and will have a strong future," Hoffmann said. "With picosecond, five years back no one used them in manufacturing. Now they are used in the production of semiconductor products."

A company that led the development of picosecond lasers for industrial microprocessing is Lumera Laser of Kaiserslautern, Germany.

While picosecond lasers have been moving into the industrial space for several years, femtosecond lasers are just beginning to make their mark.

In his plenary talk at Laser Munich, Barry Schuler, CEO and co-founder of Petaluma, Calif.-based Raydiance, said that femtosecond technology has challenges to overcome, such as the fact that it's inherently complex and unstable, expensive, and eccentric in its setup and operation but that "it has a rich potential across many applications," such as photovoltaic and microelectronic components manufacturing.

What is needed to ramp up commercialization, he said, is industrial-grade reliability, manufacturing floor robustness, ease of integration, easy maintenance and a small footprint, among other requirements. These are met in Raydiance's Smart Light, which the company says is flexible and reliable enough to be used by researchers with little or no background in optics and lasers.

"I think it's going to cause a revolution, not only in manufacturing but many other spaces," Schuler said.

A year ago, Raydiance and Rofin-Sinar Technologies Inc., of Plymouth, Mich., and Hamburg, Germany, announced that Raydiance's fiber-based laser approach would be incorporated into Rofin's StarCut Tube Femto, which allows "cold" laser cutting of medical devices such as cardiovascular stents. The machine cuts noble

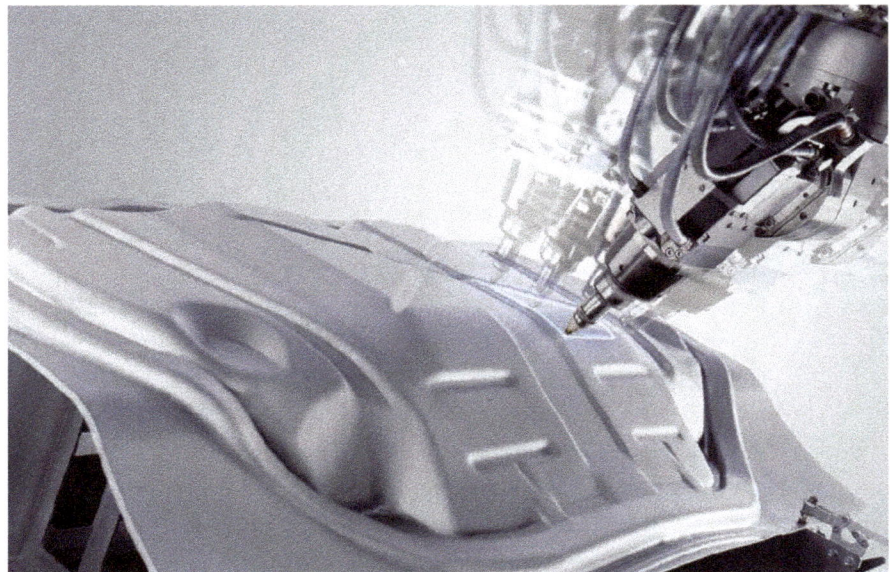

3D laser processing with the TruLaser Cell 7040. Courtesy of Trumpf Group.

metals that are difficult to machine, such as gold and platinum, as well as alloys with shape memory and polymers with a low melting point, all with no thermal effects.

Thin-disk technology also has a femtosecond laser on the market. Jenoptik Laser recently introduced its JenLas D2.fs, which the company said opens up new industrial applications for diode-pumped disk lasers due to its 400-fs pulse rate, M^2 beam quality and high pulse energies of 40 µJ at a 100-kHz repetition rate.

Diode-Pumped Lasers: Performance, Reliability Enhance Applications

The latest technology advances take diode-pumped solid-state lasers into new realms of power and wavelength, enabling many new applications.

ARND KRUEGER AND SCOTT WHITE, SPECTRA-PHYSICS, A NEWPORT COMPANY

Neodymium-doped crystals and glasses such as Nd:YAG (neodymium:yttrium aluminum garnet) have long been used as laser gain materials. Optically pumped, they produce an output wavelength close to 1 µm, and the excited-state lifetime of neodymium allows both CW and pulsed (Q-switched) operation.

The output of powerful flashlamps and arc lamps is focused into a cylindrical laser crystal rod using elliptical reflectors to form a gain module. This module is then mounted in a laser cavity, typically many inches in length and defined by the usual high reflector and partial reflector, or output coupler.

There are several limitations to this approach. First, pumping is inefficient, partly because lamps are inefficient at converting electricity into pump light and produce a lot of unwanted heat. But even more critical, the lamps produce broadband emission throughout the visible and the infrared. As a result, most of the light is not absorbed by the laser gain crystal and ultimately serves only to generate more heat in the pump module, which must be removed by water cooling of the laser head. The lamps also require a multikilowatt power supply.

For many industrial applications, the biggest drawback is the short lifetime of CW arc lamps, which must be changed every 200 to 600 hours. When the lamps are replaced, the cavity optics usually require a slight realignment to maintain

Figure 1. A flashlamp emits over a wide spectral range (b), but laser crystals such as Nd:YAG absorb light only in narrow wavelength bands (a). Diode laser pumping is efficient because the diode laser emits in only one of these bands (b). Images courtesy of Spectra-Physics.

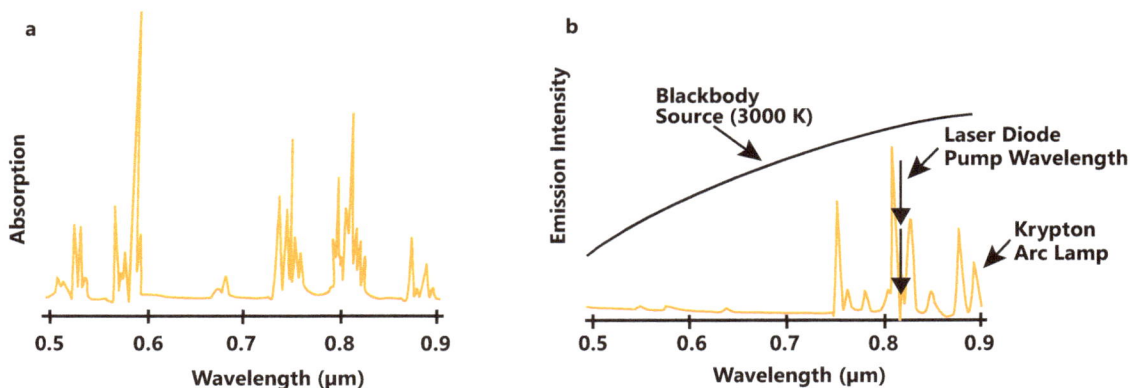

a good output mode from the laser. This frequent routine maintenance actually conceals another limitation — their optical alignment tends to drift over time and would require periodic realignment, irrespective of any lamp change. Diode pumping eliminates these limitations and drawbacks.

The principles of diode pumping are simple. The Nd-doped laser crystal has intense, sharp absorption peaks at 808 and 880 nm — wavelengths that are readily accessible to InGaAs semiconductor laser diodes (Figure 1). Laser diodes convert much of their electrical input into laser light, which is then efficiently absorbed by the Nd-doped crystal. The end result is a wall-plug efficiency many times greater than lamp-pumped lasers.

There are several other major advantages to this approach beyond electrical efficiency. Depending on the output power, these lasers generate relatively little heat and therefore may not require a high volume of cooling water like their lamp-pumped counterparts. Also, the diodes operate from a low-voltage power supply, which is compatible with a single-phase (110/220 V) line, or with lower-voltage utilities as found in some laser-based machine tools.

Furthermore, because of the compact size of the semiconductor diodes, the size of the laser head can be greatly reduced.

For OEMs and industrial end users, the long lifetime of the diodes is another advantage because it minimizes maintenance downtime. In fact, with continued advances in the reliability of diodes used in DPSS (diode-pumped solid-state) lasers, these lasers can provide many years of uninterrupted operation.

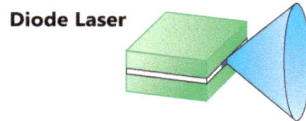

Laser geometry

There are a few basic methods for introducing diode laser pump light into a laser crystal — two of which are end and side pumping. In general, end-pumped lasers deliver high-quality output beams with state-of-the-art performance and stability at powers up to tens of watts, while side-pumped lasers sacrifice beam quality to offer raw power as high as several kilowatts.

In side pumping, laser bars or stacks are arranged cylindrically around the laser crystal (Figure 2). The output of each bar is focused by use of a cylindrical lens and/or a lens array. A large volume of the

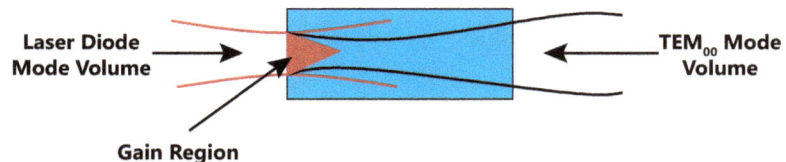

Figure 2. Side pumping enables a large number of pump bars (or stacks) to be arranged around a single laser rod.

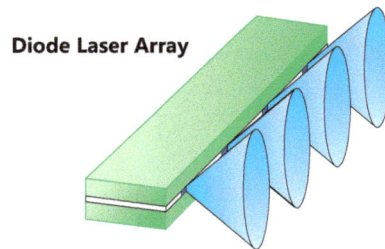

Figure 3. End pumping allows the mode volume of the diode laser to be matched to the TEM_{00} mode volume of the laser cavity.

crystal is thus flooded with pump light, leading to high power and multimode output ($M^2 > 100$). Each of these pumped crystals is mounted as a self-contained module. A high-power laser will contain multiple modules in series, each module serving to amplify the output of the previous module.

Typically, side-pumped lasers have been derived from earlier, lamp-pumped designs. Nonetheless, they offer a significant reliability advantage over traditional lamp-pumped lasers and successfully compete in heavy materials processing applications such as welding and metal cutting.

The goal of side pumping is to efficiently couple as much power as possible into the laser; in contrast, the aim of end pumping is to couple as much of the diode output as possible into the TEM_{00} mode volume of the crystal. This not only produces a lower M^2 output, but also leads to the most efficient harmonic conversion, providing access to green and UV wavelengths.

A more traditional approach for end pumping is fiber coupling, as used in the FC*bar* (fiber-coupled laser diode bar) — technology first developed by Spectra-Physics — where each diode laser facet is coupled into an individual fiber optic. The fibers are then circularly bundled such that the highly asymmetric diode bar output is converted to a high-brightness spot suitable for efficient end pumping of the laser crystal (Figure 3). Also, since the FC*bar* module(s) is mounted in the power supply and the fiber is connected to the laser head, it can simply be replaced in the field without any optical realignment.

This architecture can produce a high-quality ($M^2 < 1.2$) beam from a compact, rugged, hands-off package, with excellent maintenance lifetimes. Just as important, the flexible diode-pumped technology delivers a wide range of output powers, with a choice of CW, Q-switched and mode-locked outputs.

Enabled by innovation in laser diode technology and increased output power per emitter, a more advanced end-pumping design uses single-emitter-type diodes, which are fiber coupled directly inside the laser head, allowing for easier laser head integration and interchangeability. The life expectancy of today's diodes is more than 10 years, especially when running below their maximum-rated power (called derating). This design technique enables these models to be virtually free from diode maintenance for many years of continuous operation. Plus, a side benefit from integrating the diodes in the head is the elimination of potential issues that can arise from external fiber coupling, making laser head interchange simple and fast.

As a result of improved reliability and lower cost, end-pumped solid-state lasers have found their place in many high-precision industrial applications, dominating areas such as PCB and FlexPCB drilling and cutting, resistor trimming, stereolithography (rapid prototyping), ITO patterning, inspection, graphics, precision marking, and micromachining of a wide range of materials from glass and silicon,

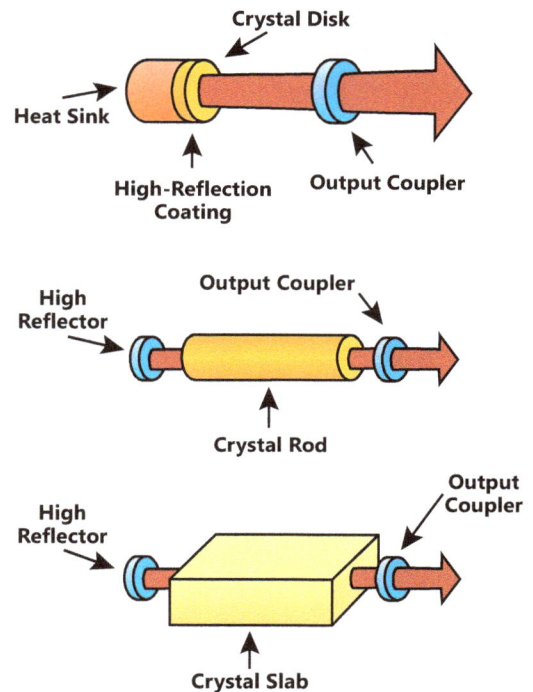

Figure 4. In commercial diode-pumped lasers, the laser crystal can be shaped as a rod, thin disk or slab.

	Nd:YVO$_4$	Nd:YAG
Table 1. **Comparison of the Important Characteristics** **of Diode-Pumped Lasers**		
Lasing Wavelength (nm)	1064.3 (π)	1064.2
Lifetime at 1 Atomic % Doping (μs)	100	220
Effective Laser Cross Section (10^{-19} cm^2)	15.6 (π)	2.8
Diode-Pumped Peak Wavelength (nm)	808.5	807.5
(π) = E⊥c		

to ceramics and metals. In addition, their low cost of ownership and operational simplicity are enabling applications that were uneconomical or impractical for earlier lasers.

Since the introduction of diode pumping, a number of laser crystal geometries have been investigated, with varying degrees of commercial success. The most important geometries are cylindrical rods, slabs and thin disks (Figure 4). Slab- and rod-shaped laser crystals can be designed to be end pumped or side pumped, depending on the power/mode requirements, while disk-shaped crystals can only be end pumped. Typically, rod-shaped crystals dominate the low-power/high-mode quality applications, whereas slabs and disks are commonly used in high-power lasers.

YAG and YVO$_4$

The most common Nd-doped material used in lamp-pumped lasers is Nd:YAG, which offers relatively simple-to-grow, large, defect-free crystals and is optically and mechanically robust. Another material, Nd:YVO$_4$ (neodymium:yttrium orthovanadate), offers higher gain than Nd:YAG. However, it was long neglected as a commercial material because the crystals were difficult to grow (i.e., pieces long enough to make laser rods for lamp-pumped systems were not available). With the advent of end-pumped configurations, the use of much smaller laser crystals is making Nd:YVO$_4$ more attractive. In addition, Nd:YVO$_4$ allows shorter-pulse-length operation, which is favorable for many applications.

In quantitative terms, Nd:YVO$_4$ has a gain about 5.5 times greater than Nd:YAG. One implication is that this allows very short pulse (<10 ns) Q-switched output with superior pulse-to-pulse stability at high repetition rates (Table 1). Nd:YVO$_4$ is strongly birefringent and naturally polarized, unlike Nd:YAG. The output of Nd:YVO$_4$ lasers is thus naturally polarized, eliminating the need for an intracavity polarizer.

With end-pumped lasers, Nd:YVO$_4$ is usually the preferred material for fast pulsed (>10 kHz) and CW operation. In fact, the maturation of this high-gain material proved critical in boosting the power of these lasers to market-enabling lev-

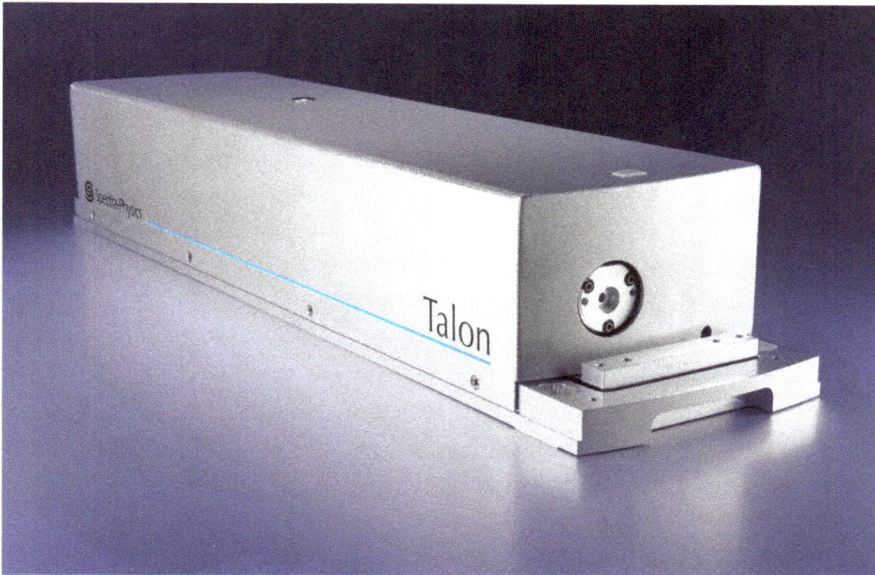

Figure 5. An example of a disruptive cost-performance laser.

els. On the other hand, Nd:YAG is still commonly used in many models operating at lower repetition rates.

The improved gain efficiency and compact designs of the newer Nd:YVO$_4$ and Nd:YAG lasers also result in significantly lower cost for performance. Figure 5 is an example of one of these newer lasers, reducing the cost/W by more than 50 percent for comparable UV and green Q-switched lasers in its class, and even lower cost than equivalent fiber lasers.

The completely sealed laser

A major reason end-pumped lasers with rod-shaped crystals are preferred in many low- to medium-power applications is that they can offer zero-maintenance operation. Operating the pump diodes derated — e.g., well below their maximum rated power — can extend the lifetimes of these devices to well beyond 100,000 hours.

This allowed a new approach to laser design, called the sealed resonator, which at the time set new standards in terms of reliability, stability, beam quality, compact packaging and simplicity of operation. For industrial applications, these advantages translate into low cost of ownership and high process yields.

Specifically, the traditional limita-

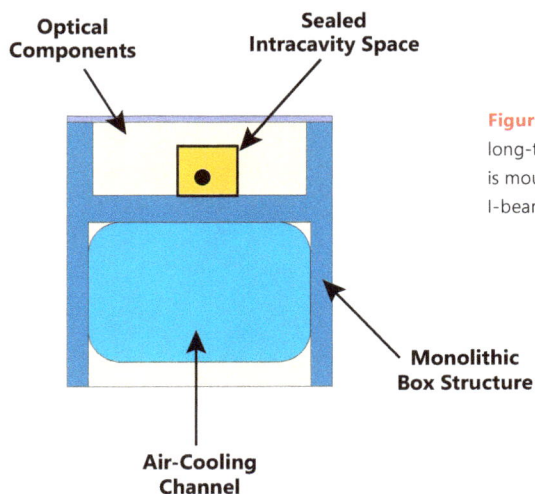

Figure 6. One method of achieving long-term stability of the laser cavity is mounting all the optics on a rigid I-beam platform.

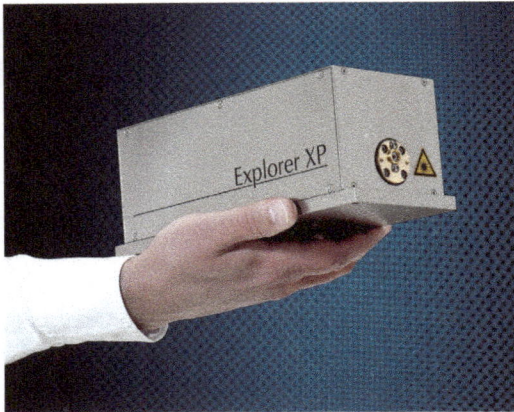

Figure 7. An example of a very compact laser.

tions of alignment drift and optical surface contamination are completely eliminated by the use of a sealed cavity. There are several successful ways to do this. Spectra-Physics uses a monolithic approach, rigidly mounting all the optics along the cross-member of a closed box I-bar structure (Figure 6). This I-bar arrangement delivers excellent torsional rigidity and stability. Also, even if the cavity temperature changes, the I-bar structure expands uniformly, ensuring that the optics stay perfectly aligned along the same axis. In addition, the optical mounts are all metal, designed for minimal use of epoxies or other outgassing components. Since there is no need to adjust or clean the cavity optics, the head is assembled and tested in a cleanroom and then sealed at the factory, eliminating optical surface contamination as a failure mechanism. Just as important, the use of small, nonadjustable mounts and fiber-coupled diodes allows for an extremely compact laser head.

Green and UV output

The near-infrared (1.06 µm) output is useful for applications such as resistor trimming and surface marking of metals, but many laser applications require visible or ultraviolet wavelengths. Fortunately, the TEM_{00} output beam produced by end-pumped lasers can be efficiently frequency doubled (to 532 nm), tripled (to 355 nm) and even quadrupled (to 266 nm) using nonlinear crystals such as LBO (lithium triborate) and BBO (barium betaborate).

With the current power level of diode-pumped CW lasers, the nonlinear crystals must be placed inside the laser cavity to obtain useful second-harmonic power levels. Spectra-Physics introduced the first multiwatt CW green laser, the Millennia, in 1996. Today, these lasers are well established with output powers exceeding 25 W.

Another route to higher green and UV power is provided by mode locking a CW laser. SBR (saturable Bragg reflector) mirror technology enables simple mode

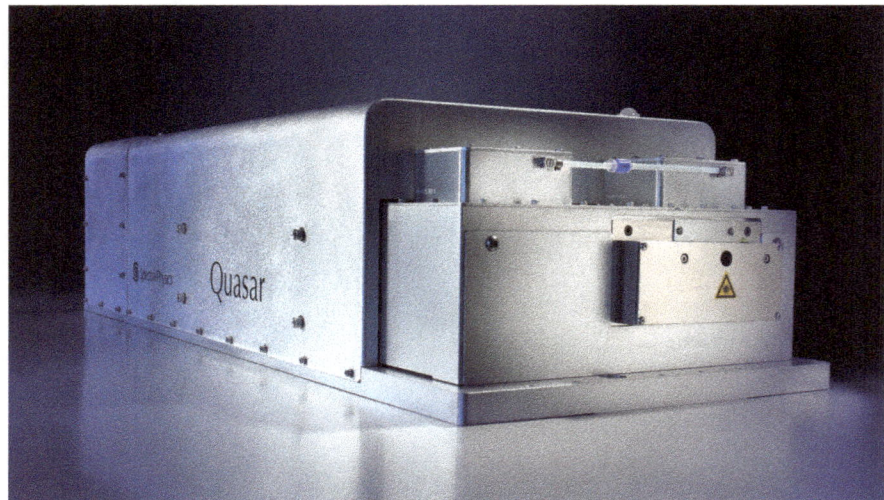

Figure 8. A hybrid fiber/DPSS laser.

locking that is robust enough for demanding industrial applications. The high peak power of mode-locked lasers allows very efficient extracavity frequency doubling and tripling, providing multiple watts of power at 355 nm. Moreover, the high repetition rate (80 MHz) means that these quasi-CW lasers have replaced bulky ion lasers in many CW ultraviolet applications.

With Q-switched pulsed lasers, the peak power is more than sufficient to permit extracavity doubling and tripling. The stable boresighting possible with a sealed monolithic laser head has allowed simple "bolt on" frequency doubling and tripling modules, as well as more cost-effective, integrated harmonic conversion methods.

Recent developments

When it comes to expanding applications for Q-switched diode-pumped lasers, there are three areas of laser development of particular note. These involve increasing the pulse repetition rate, enhancing the available green (532 nm) and UV (355 nm) output power, and manipulating laser output pulses in the time domain. Increasing the pulse repetition rate is important in industrial applications, since it leads directly to an increase in process throughput. More output power in green

Table 2.
Materials Processing Advantages with a Hybrid Fiber/DPSS Laser

Segment	Laser Processes	Materials	Quasar Advantages			
			High Power	High PRF	Short Pulse Width	TimeShift
Mobile Device Display Manufacturing	Cutting/Drilling	Glass, Sapphire	X		X	X
	TCO Patterning	ITO/Ag Film on Glass	X	X	X	X
Microelectronics Manufacturing	Laser Direct Patterning	ABF Resin on Cu	X		X	
	Via Drilling	Resin, Cu, FR4, PI	X	X	X	X
	Thru-Glass Via (TGV)	Glass	X		X	X
	Profiling	Cu, PI	X	X	X	
	Die Singulation	Silicon	X			X
	Dielectric Scribing	Low-/High-k Dielectric	X	X	X	
	Package Singulation	Epoxy Mold Compound	X	X	X	
Solar Cell Manufacturing	TFPV Scribing	TCO on Glass	X	X	X	X
	c-Si PV Dielectric Layer Ablation	SiN_x, SiO_2, Al_2O_3 on Silicon	X	X	X	X
	Metal Contact Scribing	Aluminum on Silicon	X	X		X
Industrial Manufacturing	Cutting/Drilling/Scribing	Aluminum, Stainless Steel, Coatings	X	X		
	Cleaning	Surface Oxides and Contaminants	X	X	X	X

and UV enables processing of more materials with finer precision, thus satisfying the needs for higher-density components such as mass-produced microelectronics. And tailoring the output pulses in the time domain means even more control over material interaction efficiencies to enhance processing speed and quality.

Pulse repetition rate is often the process-limiting factor in high-throughput applications such as scribing and marking. Here, the high scanning speed of the galvanometers used to sweep the laser beam cannot be fully exploited; the scan speed must be reduced to avoid producing a dotted cut or groove due to the individual laser pulses. Laser manufacturers have responded to this limitation by developing Q-switched lasers capable of much higher repetition rates.

$Nd:YVO_4$ is usually the material of choice for higher-repetition-rate lasers, but typical end-pumped designs deliver peak performance at a maximum repetition rate of only 40 to 50 kHz. Pushing these lasers to higher repetition rates often results in lower energy per pulse, lower overall power and a significant increase in pulse-to-pulse noise.

As already noted, one advantage of diode-pumped lasers is the laser head's small size. An example of a very compact laser is shown in Figure 7. The infrared, green or UV output beam is nearly diffraction limited (TEM_{00}), and allows for tight focusing and high spatial resolution. High reliability, repetition rate and pulse-to-pulse stability make this type of laser suitable for many demanding applications such as micromachining, marking and 3D stereolithography. Footprint and volume of the overall laser system can be further reduced by integrating laser head and controller into a single monolithic package.

Breakthrough technology

One of the most significant recent breakthroughs in the area of diode-pumped solid-state lasers is a hybrid fiber/DPSS laser that combines fiber laser technology and DPSS power amplification with efficient harmonic generation. One example delivering high, flexible repetition rates in the green and UV is shown in Figure 8.

Moreover, this laser exploits fiber laser flexibility not only to enable variable pulse width, but also to allow the user to control the pulse in the time domain. It produces >60 W of UV (355 nm) output power at 200 and 300 kHz and >300-µJ pulse energy, while operating over a wide range of repetition rates (single-shot to 3.5 MHz) and variable pulse widths (<2 ns to >100 ns). Green versions deliver more than 75 W at 532 nm. The hybrid architecture enables this performance in a nearly diffraction-limited beam and with very low optical noise.

The high power levels, combined with the ability to control and optimize pulse width and sequence, enable high-precision processing of a wide range of materials at very high speeds, including silicon, PCBs and ceramics. Table 2 summarizes, per material, which features are most advantageous for process quality and speed. One of the most challenging applications is the cutting of chemically strengthened glass for smartphones and tablet/PC cover plates, which can be performed at unprecedented speeds higher than 1.5 m/s.

The results indicate that process recipe development needs significant flexibility and refinement to suit the process and the user's objective. With proper parameter optimization, high quality and high throughput can be achieved, extending the capabilities of today's laser micromachining processes to meet the manufacturing challenges of tomorrow's consumer electronics products.

Conclusion

Diode pumping has revolutionized the design of solid-state lasers and enabled the creation of innovative designs to meet the evolving needs of today's manufacturing processes. The latest developments using this technology offer a unique combination of advantages, including low power consumption, low heat generation, compact packaging, excellent mode quality, high pulse-to-pulse stability, impressive high reliability and very high power at a variety of wavelengths over wide operating regimes. And by tailoring the performance of these lasers to the specific needs of new applications, laser manufacturers have ensured a healthy market for these products in a variety of existing and emerging applications.

For Lasers,
No Vibrations Allowed

*Efficient use of a laser beam is dependent not only on the power
of the laser source and beam quality, but also on its precise guidance.*

BY DR. BARBARA STUMPP, FREELANCE SCIENCE WRITER

Short- and ultrashort-pulse lasers are used to remove materials on an atomic level for high-precision applications such as extremely fine carving and cutting. Solar cells are one application where this kind of high precision is needed: Electrical contacts are placed in a small excavated area. The smaller this area is, the greater the output of the solar cell.

Lasers also are used in the manufacturing of electronics. Wherever lasers are applied, flexible handling and high resolution — as well as optimal automation of processes without long setup times — are the basis for high throughput in production. Lasers are typically surrounded by machinery and robots, systems that vibrate and oscillate, respectively — which means that lasers cannot be used in an industrial setting without a beam-guidance system. The true benefits of laser production really exist only when the beam hits exactly where it should. Where extremely fast correction and precise control are required, piezoactuators are needed. Compared with motor or inductive drive systems, piezo technology offers almost unlimited

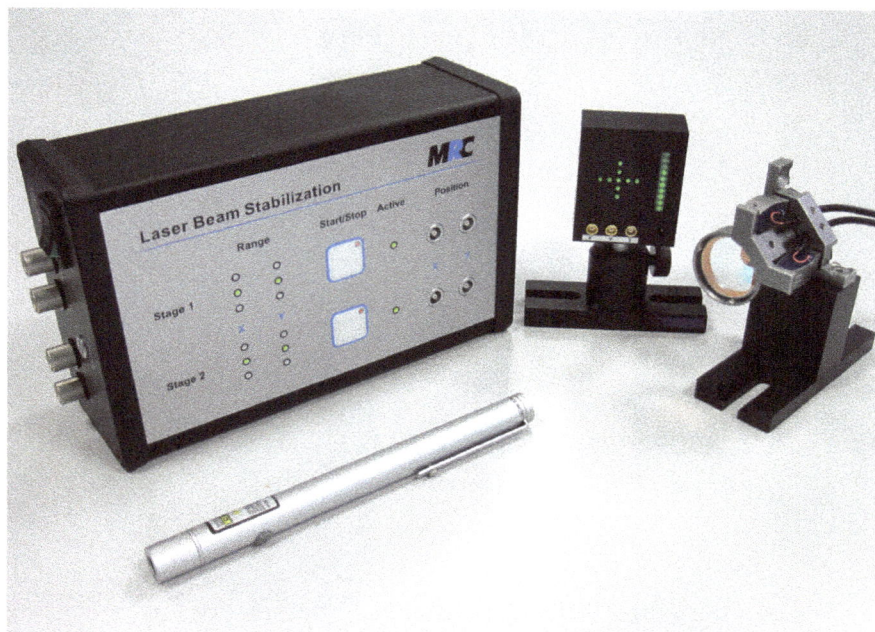

Piezoactuators enable high-precision, exact reproducibility and fast control in beam stabilization for lasers. This stabilizing system from MRC relies on piezoactuators. Courtesy of MRC.

resolution and an extremely short time delay from when the control signal is applied to when the position is reached.

Piezoactuator-based beam stabilization for lasers is characterized by high precision, exact reproducibility and fast control of the beam position. It compensates for distracting influences such as laser pointing, thermal beam fluctuations, collisions, environmental vibrations or vibrations from any other equipment.

"Actuators have to be very fast — real-time correction of the beam position is necessary, depending on the nature of the disturbance, and with an accuracy of less than 1 µm," said Dr. Marcus Goetz, CEO of MRC Systems. A real-time detector with integrated signal processing determines the fluctuation of the laser beam with highest spatial resolution. The principle even allows measurement of individual laser pulses.

The stabilization base for a laser beam is the detection of the drift with respect to the optical axis. MRC uses a four-quadrant diode (QD-4) as a detector, which can be mounted behind a highly reflective mirror and which achieves bandwidths >100 kHz. It detects deviations of the laser from the desired position and sends a signal to the controller, which transfers the signal to the tilting system — which, in turn, corrects the mirror position.

For the piezoactuator, MRC uses a mechanically prestressed, ultrafast tilting system from the PSH series made by piezosystem jena. This stage positions the mirror in two mutually orthogonal axes and an additional diagonal axis. The speed of the tilting system is very high, even under a load. The high-control bandwidth compensates for the dysfunction seen in real time. Because of the high control rate, this solution could be used with both CW and pulsed lasers. Using additional electronic

For its piezoactuator, MRC uses a mechanically prestressed, ultrafast tilting system from piezosystem jena's PSH series. Shown here, the modified design of the piezo drive for the tilting mirror. Courtesy of piezosystem jena.

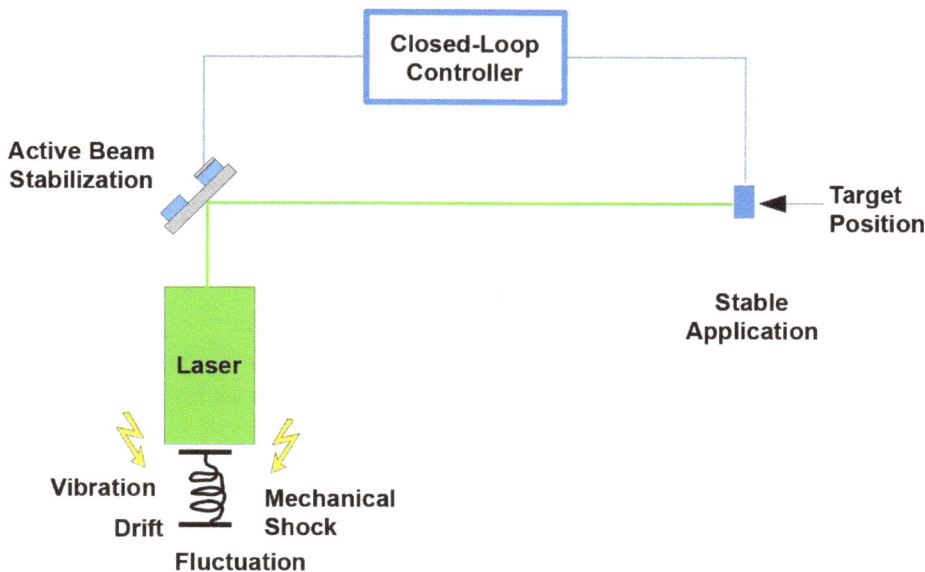

This schematic shows a working principle of laser beam stabilization. Courtesy of MRC.

circuits, even lasers with very low repetition rates or irregular pulse trains can be stabilized.

The system automatically adjusts the laser beam in a predetermined direction, counterbalancing the fluctuations in the beam direction and stabilizing the optical axis in real time. This enables active control of internal and external vibration, as well as swift and automatic correction of misalignments or pointing.

"The piezoactuators, which operate together with position detectors in a closed loop, enable fast adjustment of laser beams, to compensate [for] disturbances in the laser beam position and/or direction in real time," Goetz said. The adjustment is done over a control range of 2 mrad, corresponding with the angular displacement of the piezoactuators.

"The tilting mirror systems of our series PSH are based on the principle of the piezoelectric direct drive," said Elmar Elbinger, who is responsible for marketing at piezosystem jena. "Three separate controllable piezoactuators [tripod structure] move the centrally prestressed head plate. Therefore, these tilting mirror systems have a very high stiffness and an extremely high resonant frequency." The systems are designed for the rapid positioning tasks needed in an active beam stabilization application or beam tracking in a scanning application. Thanks to very low settling time, any angle constellations can be approached dynamically. Systems with integrated position sensors also allow for precise repeatable control of defined positions.

A special feature is the temperature-compensated design of these tilting mirror systems. The piezoactuators are also mechanically prestressed and, therefore, particularly suitable for dynamic applications. They also are available for use in vacuum and at low temperatures.

Slight footprint

With laser-stabilization applications, real-time capability is important, but so is compact construction — which is true for many industrial systems. MRC required the placement of the mirror to be far away from its tilt axis, but this would cause too large moments of inertia, causing problems for the actuators. An alternative would have been additional optical elements in the beam path; however, this would not have improved the beam quality. The developers at piezosystem jena found a simple solution: adapting an existing design. This was also the most economical solution: Standard components could be used, and the assembly position of the PSH in the MRC system could be accomplished horizontally.

By introducing an appropriate notch in the actuator, the mirror could stay close to the tilt axis, and the torque required to tilt the mirror would be sufficiently low. Therefore, the proven reliability of the drive concept was guaranteed even during continuous use, and the dynamic parameters of the PSH system remained.

"The horizontal assembly position of the piezoactuator is parallel to the beam path. Thus, there is no beam deflection needed in the system," Elbinger said. "The PSH — with a resonance frequency of 5.4 kHz — can be used up to 80 percent of its resonance, which means the piezo-system can work up to a frequency of 4.3 kHz. With a high frequency, dysfunctions that appear in this range can be eliminated. The high resonant frequency makes the system less susceptible to external oscillations.

"Also, low-frequency dysfunctions have less of an impact on piezos with high resonance. Lastly, disturbing oscillations even in the higher frequency range can be eliminated."

The quality of the stabilization is defined by the speed and precision of the control. Without the piezoelectric drive elements working in real time, the following processes could be neither optimized nor realized: thermal stabilization of a femtosecond laser, high-resolution position correction, fast position correction of a laser, vibrations in lithography or scanning 100 lines of an area in less than a second. For all these applications, laser-beam stabilization is needed to maintain performance and to increase scope.

Meet the author

Dr. Barbara Stumpp works as a freelance journalist in Europe and does public relations work for enterprises from the business-to-business areas. A version of this article appeared in German in *Laser+Photonik* in May 2013.

An Introduction to Laser Process Qualification

The step from successfully demonstrating a new laser process on a limited number of samples under laboratory conditions, to scaling up that process for high-volume industrial use is a big one. In particular, it's necessary to determine how variations in process parameters affect end results, so that acceptable process input and operating ranges can be determined.

BY JOCHEN DEILE AND FRANK GAEBLER, COHERENT INC.

Materials processing with lasers is actually a broad term for a diverse range of applications. The major categories of these uses, and the lasers that service them, are summarized in the table on page 240. At one end, multikilowatt fiber and CO_2 lasers are employed for cutting and welding metal in industries such as automotive, shipbuilding and appliance production. At the other end of the spectrum, ultraviolet and ultrafast lasers are used to drill micron-sized holes and scribe thin films in high precision processes for microelectronics, display and solar cell fabrication.

Process qualification is important in all of these areas, although the main focus and primary considerations change across the spectrum of applications. For example, in many sheet-metal cutting applications, the environments are not well-controlled, and there are often batch-to-batch variations in incoming materials. In

Figure 1. Different types of materials processing, such as cutting and welding metal or drilling micron-sized holes and scribing thin films, call for different types of lasers. Whatever laser is chosen to do the job, the statistical discipline of process qualification is crucial to ensuring a process can operate successfully and around the clock despite variations, such as materials, fixtures, laser output, beam delivery and ambient conditions.

this case, it's important to develop a process that can operate successfully despite these variations, often for automated, around-the-clock use with minimal operator intervention.

In contrast, the manufacturing environments for high precision processes are typically tightly controlled. But, the need to hold tighter mechanical tolerances may put greater emphasis on the development of precision fixtures. Ultrafast processes in particular may require better control of optical alignment in order to maintain the high peak powers needed to drive nonlinear effects.

Cost is key

Whatever the particulars of the fabrication task, the first input for the process qualification procedure is a statement from the manufacturer of what the process must achieve. Specifically, this involves setting nominal values and tolerances for all process results, such as the diameter, depth and position of a laser-drilled hole.

The next key piece of input for process qualification is cost. For most manufacturers, the actual metric is cost per part. This input must be considered at the very outset, because any new production methodology will only be adopted when it is cheaper than the existing method for accomplishing the same task; when it offers significantly greater operational flexibility; or when it allows producers to increase the functionality or value of their products in a way that enables them to raise their own price to the consumer.

An overview of the major categories of materials processing applications, and the lasers that service them.

Process	Laser Type					
	DPSS Pulsed (ns)	DPSS Pulsed (Ultrafast)	CO$_2$	Excimer	Fiber	Direct Diode
Thin-Film Scribing	●	●	●			
Silicon Annealing				●		
Cutting	●	●	●		●	
Converting	●		●			
Drilling	●	●	●	●	●	
Marking & Engraving	●	●	●	●	●	
Welding			●		●	●
Cladding/Heat Treating					●	●
Rapid Prototyping	●		●		●	
Surface Treatment		●		●		

Coherent Inc.

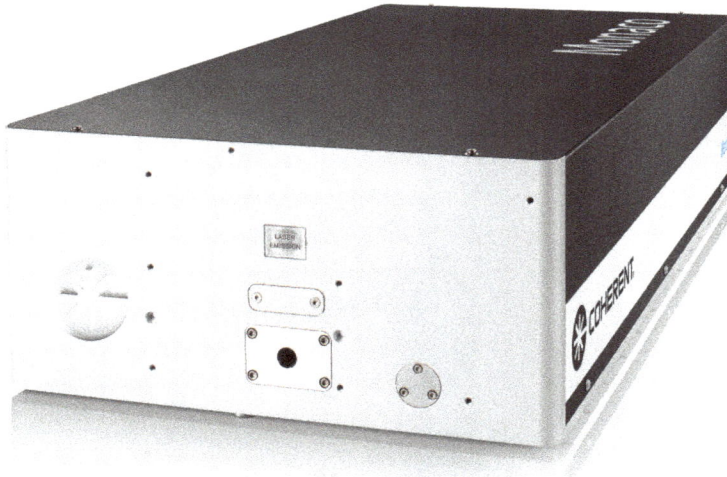

Figure 2. Industrial ultrafast lasers, such as the Coherent Inc. Monaco (400-fs pulsewidth), are becoming increasingly popular for high precision materials processing. Their high peak power and short pulsewidth produces reduced heat affected zones (HAZ) and a lower ablation threshold as compared to longer pulsewidths, thus enabling high precision processing of a wide range of metal, semiconductor and organic materials. Courtesy of Coherent Inc.

There are several factors that influence cost per part. The most obvious are the capital cost of the production equipment and its cost of ownership characteristics, such as power consumption, other consumables and maintenance costs. However, there are many other factors, as well. For example, the production rate, meaning the number of parts produced in a given time period, is usually a significant consideration, especially if the process represents a production bottleneck. Also, the incremental capacity of the machine may be a consideration. Specifically, would it be better from a risk mediation point of view to utilize several lower-capacity machines, even though they are less cost-effective? Process yields, the level of rework required or the need for some form of post processing, such as part cleaning, also directly impact cost.

Another significant cost issue is the manpower required to support it. Does the equipment require an operator, and, if so, at what skill level? Can the new machine be readily integrated with other production equipment? If necessary, does it lend itself to automated part loading and unloading?

The maintenance requirements for a machine may also affect the manpower budget, specifically if the manufacturer needs to have specialized personnel on staff to perform routine maintenance and adjustment of the system. This is a particular consideration for laser-based equipment, because laser and optics adjustment may be outside of the expertise of a company's existing maintenance staff. Of course downtime, either for routine maintenance or repair, plus the cost of repair itself, can also have a major cost impact.

Process variables

Once targets for all these factors are determined, then process qualification involves developing a production methodology that meets these cost and quality constraints, and that can be reproduced on a large scale basis. Process qualification is, by its very nature, a statistical discipline. It is not sufficient to simply prove a given process on a small number of parts, because this will fail to show

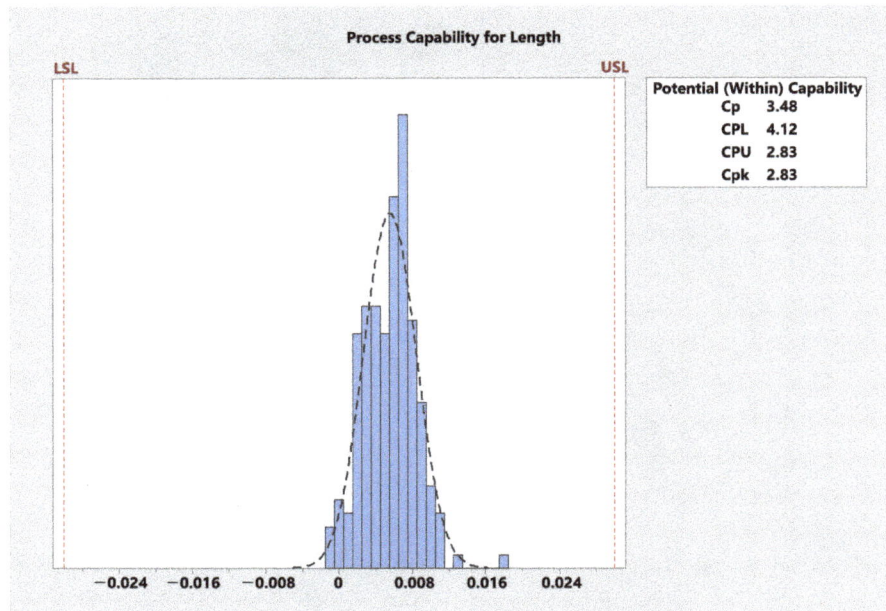

Figure 3. Normalized statistical data acquired for length of the processed glass parts. Courtesy of Coherent Inc.

the underlying variations that will naturally occur when the process is scaled up to higher quantities. Thus, enough testing must be performed to yield a large enough sample size so that accurate statistics can be determined on the level of part-to-part variation that can be expected in volume production.

Several factors lead to the variation in process output just mentioned. Some of the most obvious ones include materials, fixtures, laser output, beam delivery, ambient conditions and other process input conditions, such as electrical power or cooling water supply. It's worth examining these factors in more detail.

In a real world manufacturing environment, materials are often sourced from multiple vendors, and even within an individual vendor's product, there may be batch-to-batch or unit-to-unit inconsistencies. These variations can be broadly placed into two classes. The first is dimensional variations, such as thickness, which obviously impact processes such as cutting and drilling.

The second type of material variation is intrinsic, or differences in the properties of the material itself, such as absorption or reflectance characteristics, refractive index, thermal conductivity, internal stress characteristics, chemical composition and the like. These variations may directly affect the laser-material interaction, and alter process results.

In terms of laser output, the most obvious variable that affects process outcome is laser power or pulse energy. Furthermore, many processes are sensitive to pulse shape, especially to the extent that it determines peak power. This is particularly critical for ultrafast lasers, which often interact with materials through a nonlinear process, which is necessarily highly dependent upon peak power.

Changes in ambient environmental conditions, specifically temperature and humidity, can impact a process in several ways. For example, both optomechanical mounts and optics themselves may respond to temperature changes in a way that alters beam pointing or the beam shape. This changes the geometry or position of the focus, which becomes more critical when producing smaller features. Again, the size, and hence fluence on target, is especially critical with ultrafast lasers since the ablation rate is a function of fluence.

Even in the absence of any environmentally produced fluctuations, there is always a variation in the performance of mechanical parts. For example, if parts are loaded into a given mechanical fixture 1,000 times, there is going to be a distribution in the position of where the part actually sits each time.

Real world process qualification

In theory, the ideal way to determine acceptable limits for each process variable would be to vary each parameter individually while holding all the others constant, and then see the impact of this change on results.

However, this approach is impractical, and, usually, impossible. This is because there are typically a large number of variables present in a given process, and some of them, such as laser pulse shape, cannot easily be varied deterministically.

As a result, real world process qualification usually starts with system design. Specifically, this involves selecting system components, such as lasers, optical mounts and part fixtures, which have already proven themselves to be stable over time under relevant operating conditions. In other words, picking a highly stable laser source eliminates the need to test the process for the effects of variations in laser power or pulse-to-pulse stability.

Also, in some cases, it's not necessary to test the influence of a given variable. Instead, some overhead is simply added into the process so that it will work properly, even in the presence of the expected variation. For example, it might be determined during testing that, under ideal conditions, a laser cutting process could

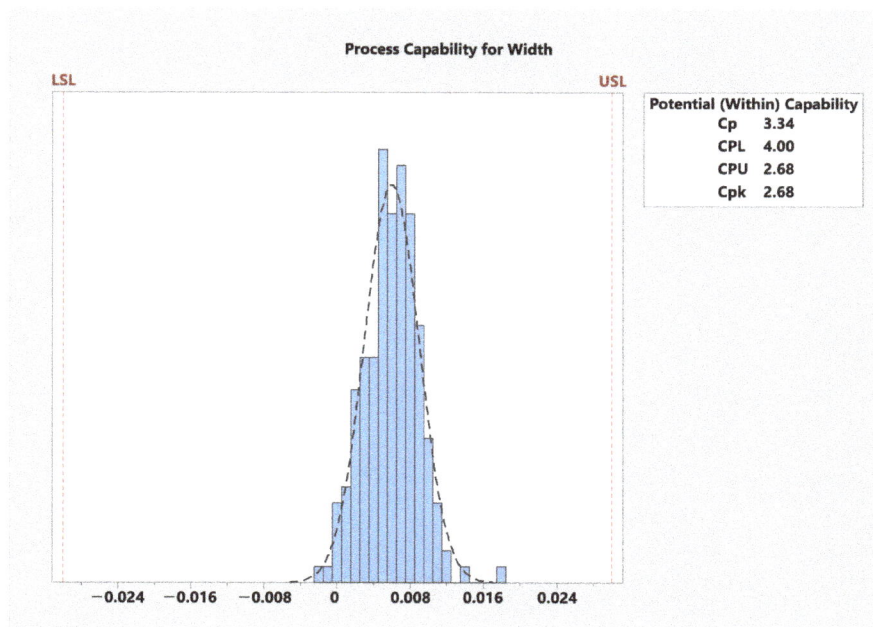

Process Capability for Width

Potential (Within) Capability	
Cp	3.34
CPL	4.00
CPU	2.68
Cpk	2.68

Figure 4. Normalized statistical data acquired for width of the processed glass parts. Courtesy of Coherent Inc.

operate successfully at a feed rate of 1 m/s. But, since it is known that there will be material variations in production, the production feedrate is set to 0.8 m/s to accommodate for this.

Of course, this has a cost impact. Running a process at a slower rate reduces production, but it will probably also lower scrap rates. Because of this, a key aspect of process qualification is identifying which factors are most critical. Specifically, what outcomes are absolutely necessary to proper operation of the finished part, and which parameters are less critical.

Finally, a sufficient quantity of parts must be produced using the process under test in order to provide a statistically valid determination of the levels of part-to-part variation that can be expected in volume production.

Display glass application

Reviewing an actual process qualification example might help to make this all more concrete. In this case, Coherent Inc. was contracted by a display manufacturer to qualify a process for cutting thin glass — less than 1 mm — using an ultrafast laser.

This producer had been cutting glass mechanically, but this necessitated grinding and polishing post-cut to deliver the desired cut edge quality. Also, mechanical cutting didn't have the capability to produce curved cuts, or to work with non-flat glass, both of which were becoming requirements in this application. Ultrafast laser processing offered the potential to overcome all these drawbacks. The producer determined the production rates, tolerances and yields that would be necessary in order for the ultrafast laser-based process to be a cost-effective replacement technology.

Figure 5. Normalized statistical data acquired for edge roughness of the processed glass parts. Courtesy of Coherent Inc.

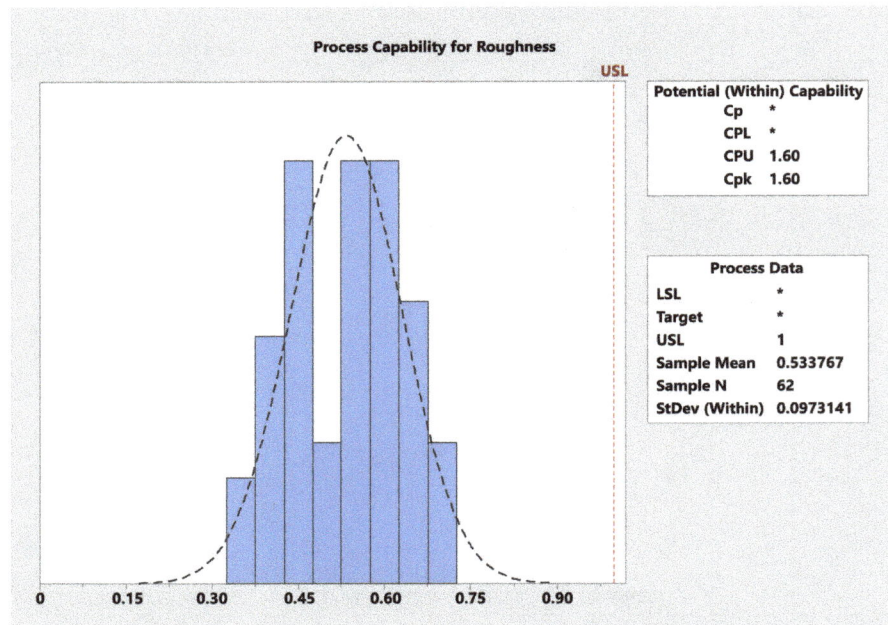

Process Capability for Roughness

Potential (Within) Capability	
Cp	*
CPL	*
CPU	1.60
Cpk	1.60

Process Data	
LSL	*
Target	*
USL	1
Sample Mean	0.533767
Sample N	62
StDev (Within)	0.0973141

It was quickly determined that a workstation employing high-speed, motorized XY stages for part positioning would be required in order to meet the processing time requirements. The task then became determining a compromise between acceleration values for stage movement that minimized processing time, while still producing positioning errors that were within the user's specifications for dimensional accuracy. Then, the goal was to find the most cost-effective controller and stage package that could reliably deliver this level of performance.

A series of test runs using various stage acceleration values were performed to determine these tradeoffs. These tests were repeated three days in a row, several times a day, to prove that the process was stable over time and as ambient operating conditions varied. Testing also utilized parts from several lots to account for batch-to-batch differences in the material. In particular, variations in internal stress in the glass had already been identified by the customer as a problem that affected end results.

After processing, parts dimensions were measured using an optical microscope. Plus, an optical surface profilometer was employed to gauge surface quality. From this, statistical plots of the results were assembled for the distribution in the length and width dimensions of the cut piece and in the processed edge surface roughness (Figures 3-5). The length and width plots are histograms of the particular dimension measured, along with the upper specification limit (USL) and lower specification limit (LSL) provided by the manufacturer. Plus, a best fit line to the distribution of values measured is shown.

From this data, the statistical quantity for the process capability (Cpk) was calculated. Cpk is a number that quantifies how closely a process is performing to its specification limits, relative to the natural variability of the process. The higher the value for Cpk, the less likely it is that any given part produced will be outside of the specification limit. For example, a Cpk value of 1.33 represents a process yield of 99.99 percent.

In this example, the process was qualified by determining the Cpk values for the mechanical tolerances in three runs in over three consecutive days. The plots show the capability for the combined data. More specifically, they plot deviation from nominal, not the actual value for the dimensions. With Cpk values of well over 2, the process as it was configured was proven to be highly capable. In addition, the surface roughness was measured, and a Cpk value of 1.6 was calculated. Again, this showed the process to be operating extremely well.

In conclusion, process qualification is a necessary step in moving from the "proof of concept" stage into production reality. Because the nuances of laser processing are not always well-understood by end users, this can often be best accomplished by working with a laser supplier that has capabilities for performing process validation studies.

Meet the authors

Jochen Deile is a product manager for Coherent Inc. in Richmond, Calif. Frank Gaebler is the director of marketing for Coherent Inc. in Dieburg, Germany.

Engineering Makes Powerful Lasers Safer

Containment, software, training and other precautions allow embedded or enclosed lasers to move into new applications.

BY HANK HOGAN, CONTRIBUTING EDITOR

Hiding your light under a bushel isn't always a bad thing, contrary to the advice in the old proverb. A case in point can be found in lasers, where low-cost systems are increasingly more powerful — and, therefore, potentially more dangerous.

Using techniques and technology that effectively hide light away, however, these powerful lasers can be embedded in systems safe for use in the open by the general public. Although safe, embedded lasers face challenges as they run up against the drive to make tools that are smaller, lighter and more flexible.

Classification, containment

Lasers are categorized according to their ability to do damage, with Class 1 products considered safe at all times under all normal-use conditions. At the other extreme, Class 4 products can burn the skin, blind eyes or ignite materials. But, with engineering, even dangerous lasers can be rendered benign.

With engineering, lasers powerful enough to drill through metal during fabrication of aerospace parts can be used safely. Courtesy of IPG Photonics.

"Classification is based on human access to laser radiation during operation of the product," said Jay Parkinson, a certified laser safety officer and president of Phoenix Laser Safety LLC. "CD/DVD players have embedded Class 3B or 4 lasers but are Class 1 laser products."

This feat is pulled off by ensuring that the more hazardous levels of light from the beam cannot leave the product housing. In general, the class of an embedded laser will drop through the use of protective coverings, as well as safety interlocks, that shut the beam off if the housing is opened. There also may be carefully placed viewing windows and specially designed optics. A final tool in the safety arsenal is the use of scan-failure safeguards. These ensure that the beam doesn't stop moving due to a breakdown in the mechanism that moves it from place to place. For example, scan-failure safeguards can keep a beam from dwelling too long on spots not designed to handle a stationary laser.

IPG Photonics Inc. of Oxford, Mass., makes fiber lasers from milliwatt output up to systems topping out in the tens-of-kilowatts range. The company has been designing and supplying laser-based systems for more than 20 years.

Its Microsystems Div. makes workstations for machining holes and other small features. These systems are Class 1 or Class 4, which can be considered closed- or open-beam. While the first category can operate safely in open spaces, the second requires safety precautions such as beam containment in temporary enclosures, warnings, interlocks and operation by trained personnel equipped with personal safety gear.

Such requirements can be a burden to those running the laser, but may be unavoidable. Manufacturers, after all, have throughput, quality and cost targets to hit, and these targets dictate laser characteristics.

"In general, the laser process is set by the machining requirements, and the workstation is designed to enclose the light and protect personnel. It is very rare that a process would deviate from an optimal manufacturing process rather than engineer the proper safety enclosures, interlocks and beam handling devices," said John Bickley, the Microsystem Division's director of sales and marketing.

The power in a beam can be substantial, and the future promises even more powerful lasers — and this trend can be seen in the offerings from laser compa-

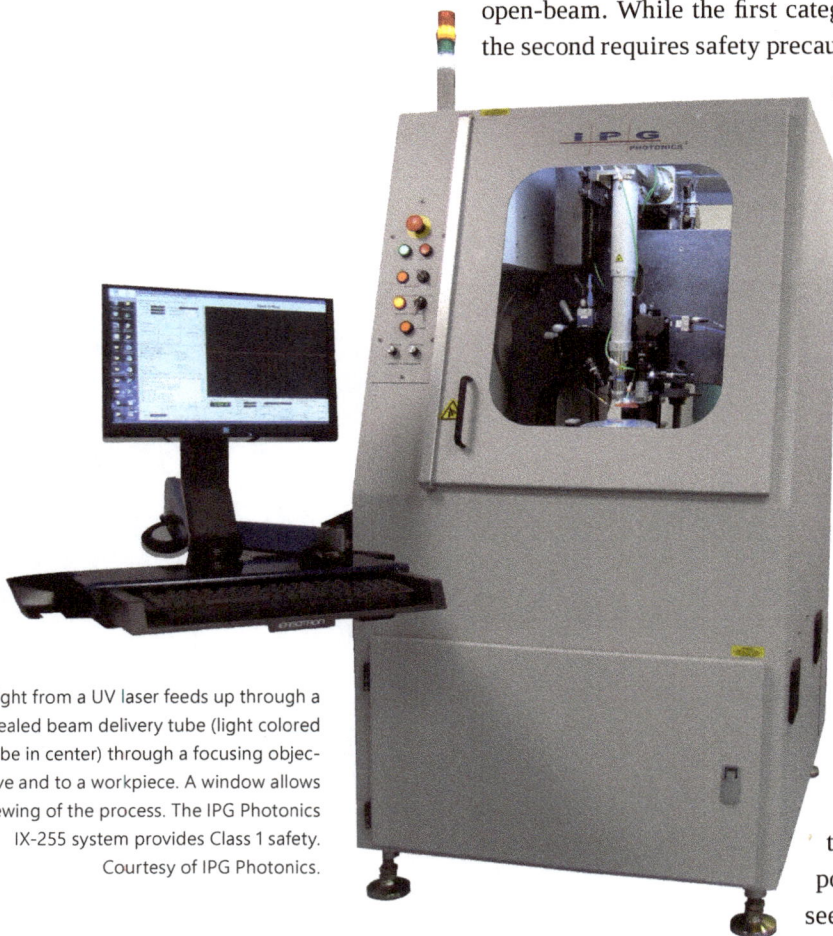

Light from a UV laser feeds up through a sealed beam delivery tube (light colored tube in center) through a focusing objective and to a workpiece. A window allows viewing of the process. The IPG Photonics IX-255 system provides Class 1 safety. Courtesy of IPG Photonics.

nies. IPG Photonics, for instance, has supplied commercial lasers up to 100-kW output.

The effect of growing the power of enclosed lasers may be particularly evident in metal cutting and welding. These and other power-intensive applications may have multikilowatt lasers mounted on robot arms, which present a particular challenge.

"As those things go up in power and complexity, and you have longer and longer focal lengths, you have virtual light sabers on the ends of robots. It can get pretty complex to contain them," said Thomas J. Lieb, certified laser safety officer and president of Elk Grove, Calif.-based L-A-I International.

Software, wavelength, pulse width

Although advances in technology have made some containment more challenging, other innovations have made the job easier; e.g., in times past, laser safety experts had to worry about robots malfunctioning, with one possibility being that the beam would end up stationary, or almost so, on a containing wall. One solution to such a case would be multiple walls, under the theory that the extra protection would afford time to notice the error and shut down the machine.

In the past few years, however, licensing safety bodies have approved safety software, which enables protections that prevent this from happening. The software also keeps maintenance or other personnel from overriding the system and thereby forcing it to act in unintended ways.

Other trends that affect safety have to do with wavelength. A decade or so ago, according to Lieb, industrial applications were dominated by CO_2 lasers operating in the far-IR, well outside the retinal hazard region. Now, more powerful beams are showing up in the near-IR, say from 700 to 1400 nm. Such light is invisible but presents a burn hazard; also, eye damage is possible from low-level reflections. Consequently, people exposed to such lasers may end up with a dead spot in an eye without ever feeling a thing. That puts a premium on making sure that those working with or even strolling by such systems cannot encounter the beam.

That's one reason why major automakers and their suppliers insist that all laser systems as delivered be Class 1 devices. That requirement is a consequence of applying current standards, when the applicable documents are read in total, Lieb noted.

Industrial lasers also are operating at shorter pulse widths. Doing so can yield

A robot holds a beam end-effector, with the laser (not shown) making welds. Rendering safe high-power lasers in systems with long focal lengths is challenging, particularly when such systems are mounted on robot arms. Courtesy of L-A-I International.

more precise cuts with much more limited damage to surrounding material. That and the lower cost of the technology are some of the reasons why there has been a move of industrial lasers into the picosecond- or femtosecond-pulse-width regimes.

Exposure levels

These technology changes are taking place while the acceptable exposure level is being reconsidered. Dr. David H. Sliney is now a consultant, but was formerly the program manager for laser/optical radiation at a U.S. Army Center at the Aberdeen Proving Ground in Maryland. He is currently shepherding a revision of the applicable standard, IEC 60825-1, through the approval process. It should be published soon, he said.

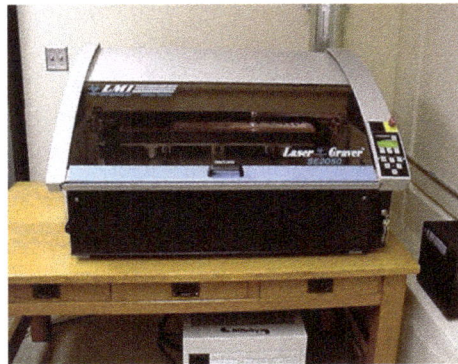

A laser marker with an embedded Class 4 laser inside is a Class 1 system, thanks to engineered safeguards. Courtesy of Laser Institute of America.

The update has changes that affect embedded lasers and safety considerations. For instance, the new revision changes how repetitive pulses are tabulated in terms of exposure.

"We recognized in the laser bioeffects research that this rule that's been around 20 or 30 years really was an error," Sliney said. "It appeared that the statistics were showing that the hazard increased with many pulses. In fact, it does not."

As a result, the allowable limits will be going up for repetitive pulses. Similarly, the limits for the shortest pulses, those that range from 100 fs to several picoseconds, are going up. Biophysics research is also behind a relaxation of the limits for the highest power pulses, according to Sliney.

Not every class of pulses will see a loosening of exposure limits. The allowable radiation exposure from single nanosecond-length pulses is going down, for example.

Industry-specific changes

There also are some changes that are related to specific products and industries. For instance, the movie industry is moving from xenon arc lamps to laser illumination. The way the standard was written, that transition required the use of ropes to keep moviegoers and others away, as well as a laser safety officer. The new light source, however, is no more dangerous than the old, according to the industry. Consequently, the new standard has specific provisions to eliminate the most burdensome laser safety requirements for projection applications.

As can be seen by such an example, setting the safety parameters for an embedded laser system is a complicated affair. Likewise, meeting the requirements can be a knotty challenge. It's made more difficult because the safety systems add cost and complexity, which may make a laser-based solution less attractive than the alternatives.

Gus Anibarro, the education director of the Laser Institute of America, teaches the organization's industrial laser safety officer (LSO) course and also a three-day LSO course. During this training, he goes over enclosures. The most popular in an industrial setting are made of steel; in research facilities, the enclosure is most often made of a combination of polycarbonate and brush, black anodized metal.

Containers can ensure no laser radiation escapes and so eliminate any chance of harm. Such an arrangement makes an otherwise dangerous laser harmless and allows it to be widespread.

A good example is a laser printer, Anibarro said. "Typically, these lasers are Class 3B. You can pull out the ink cartridge, pull out the paper bin, and you will never be exposed to the laser beam. This is a Class 1 laser system, and you probably have one in your office."

Dictionary

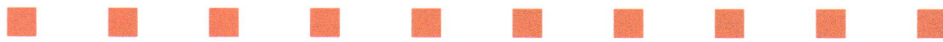

Dictionary

A

ablation threshold
The minimum energy required to induce atomic and molecular separation or displacement due to incident intense laser irradiation.

absorption
The transfer of energy from an incident electromagnetic energy field with wavelength or frequency to an atomic or molecular medium.

absorption spectrum
Fraction absorption over a specified range of wavelengths.

actuator
Mechanical device intended for the translation (rotational and linear) using high precision control from electronically operated circuits.

adhesion
The intermolecular attraction between two surfaces, as between a substrate and a coating; it is an important factor in the durability of optical thin films.

amplifier
A device that enlarges and strengthens a signal's output without significantly distorting its original waveshape. There are amplifiers for acoustical, optical and electronic signals.

analyzer
An optical device, such as a Nicol prism, capable of producing plane-polarized light, and used for detecting the effect of the object on plane-polarized light produced by the polarizer.

angular misalignment
Angular deviation from the optimum alignment of source to optical waveguide, waveguide to waveguide, or waveguide to detector, resulting in a loss of optical power.

anisotropic
Describing a substance that exhibits different properties along different axes of propagation or for different polarizations of a traveling wave.

annealing
The process of heating and slowly cooling a solid material, like glass or metal, to stabilize its thermal, electrical or optical properties or, as in semiconductor materials, to reverse lattice damage resulting from ion implantation of dopants.

aperture
An opening or hole through which radiation or matter may pass.

assist gas
A gas, such as oxygen, that improves the speed and efficiency of a laser cutter or welder when applied to the work surface, or an inert gas, such as argon, that helps to clean or shield the work surface.

average power
In a pulsed laser, the pulse energy in joules times the repetition rate in hertz.

axis
A straight line, real or imaginary, passing through a body and indicating its center; a line so positioned that various portions of an object are located symmetrically in relation to the line.

B

backlit
Refers to a display or screen that is illuminated from behind; the light is transmitted as opposed to reflected.

backscatter
The deflection of radiation by scattering processes through angles that exceed 90° with respect to the original direction of motion.

bandgap
In a semiconductor material, the minimum energy necessary for an electron to transfer from the valence band into the conduction band, where it moves more freely.

baseline
The smallest amount of photon energy to pass a detector window and be counted.

• For a comprehensive dictionary of optics and photonics terms, please visit photonics.com/EDU.

---

beam
1. A bundle of light rays that may be parallel, converging or diverging. **2.** A concentrated, unidirectional stream of particles. **3.** A concentrated, unidirectional flow of electromagnetic waves.

beam diameter
1. Calculated distance between two exactly opposed points on a beam at a chosen fraction of peak power (typically $1/e^2$). **2.** The diameter of a circular aperture that will pass a specified percentage (usually 90) of the total beam energy.

beam divergence
Increase in the diameter of an initially collimated beam, as measured in milliradians (mrad) at specified points; i.e., where irradiance is a given fraction (often $1/e^2$) of peak irradiance.

beam expander
A system of optical components designed to increase the diameter of a radiation beam. Usually an afocal system.

beam position
In computer graphics, the point on the display screen where the electron beam is located before the display instruction is executed. On directed beam display points, vectors and other graphic elements are often defined in relation to the current position of the beam.

beam profiler
A device that measures the spatial distribution of energy perpendicular to the propagation path of a radiant beam. An energy or power meter is typically used to monitor the amount of light passing through a slit, pinhole or aperture that is scanned across the beam path.

beam waist
That point in a Gaussian beam where the wavefront has a curvature of zero and the beam diameter is a minimum.

beamsplitter
An optical device for dividing a beam into two or more separate beams. A simple beamsplitter may be a very thin sheet of glass inserted in the beam at an angle to divert a portion of the beam in a different direction. A more sophisticated type consists of two right-angle prisms cemented together at their hypotenuse faces. The cemented face of one prism is coated, before cementing, with a metallic or dielectric layer having the desired reflecting properties, both in the percentage of reflection and the desired color. In a color television camera, for example, a three-way beamsplitting prism is employed in which multilayer films are deposited on the interfaces to divert red and green light to two vidicons, leaving the blue image to pass through to the third vidicon tube.

birefringence
The separation of a light beam, as it penetrates a doubly refracting object, into two diverging beams, commonly known as ordinary and extraordinary beams.

bore
The central hole running the full length of a laser capillary tube, in which electrical discharge and laser action take place. Also, a similar hole in a hollow waveguide or a microchannel plate.

borosilicate glass
A strong, heat-resistant glass that contains a minimum of 5 percent boric oxide.

Bragg grating
A filter that separates light into many colors via Bragg's law. Generally refers to a fiber Bragg grating used in optical communications to separate wavelengths.

C
capacitance
The ability of a conductor to store an electrical charge; its value is given in farads as the ratio of the stored charge on one conductor to the potential difference between it and a second conductor.

carbon dioxide (CO_2) laser
A gas laser in which the energy-state transitions between vibrational and rotational states of CO_2 molecules give emission at long IR, about 10 µm, wavelengths. The laser can maintain continuous and very high levels of power.

cavity
In a laser, the optical resonator formed by two coaxial mirrors, one totally and one partially reflective, positioned so that laser oscillations occur.

cavity dumping
A Q-switch method that can result in extreme pulse shortening.

chip
1. A localized fracture at the end of a cleaved optical fiber or on a glass surface. **2.** An integrated circuit.

cladding
The low-refractive-index material that surrounds the core of an optical fiber to contain core light while protecting against surface contaminant scattering. In all-glass fibers, the cladding is glass. In plastic-clad silica fibers, the plastic cladding also may serve as the coating.

collimator
An optical instrument consisting of a well-corrected objective lens or mirror with a light source and or object/

image (i.e. illuminated slit or retical) at its focal plane. Collimators are used to calibrate and align optical devices and elements, determine focal lengths, as well as replicate and project an source/object or image to infinity.

component

1. A constituent part. It may consist of two or more parts cemented together, or with near and approximately matching surfaces. **2.** The projection of a vector on a certain coordinate axis or along a particular direction. **3.** In a lens system, one or more elements treated as a unit. **4.** An optical element within a system.

compound lens

A lens composed of two or more separate elements of optical glass that may or may not be cemented together. The surfaces of the elements are shaped to reduce or eliminate the aberrations inherent in a single lens.

compound semiconductor

A semiconductor made up of two or more elements, in contrast to those composed of a single element such as germanium or silicon. In a III-V semiconductor, for example, one or more elements having three valence electrons (gallium, for instance) are combined with one or more having five (arsenic).

concave

Describing a hollow curved surface; curved inward.

connector

Hardware installed on fiber cable ends to provide cable attachment to a transmitter, receiver or other cable. Usually a device that can be connected and disconnected repeatedly.

convex

Denoting a spherically shaped surface; curved outward.

cross section

Calculation of the probability of an interaction between two types of particles, such as light absorption, excitation or energy transfers. The probability that one incident particle will interact as it passes through the target layer is defined by the cross section multiplied by the number of target particles per unit area.

D

deflection

Any bending of a wave of radiation away from its expected path, as, for example, by diffraction or by a magnetic field.

degradation

The gradual decrease over time in output signal with constant input light level.

degrees of freedom

The number of unique ways in which a part can move in an alignment system. In static alignment, there are six: one in the direction of and one in rotation about each of the X, Y and Z axes. In dynamic alignment, as of a scanning system, there are up to five more. When aligning a part, each degree of freedom must be constrained sufficiently to avoid unwanted motion but not so much as to create stress.

detector

1. A device designed to convert the energy of incident radiation into another form for the determination of the presence of the radiation. The device may function by electrical, photographic or visual means. **2.** A device that provides an electric output that is a useful measure of the radiation that is incident on the device.

diffraction

As a wavefront of light passes by an opaque edge or through an opening, secondary weaker wavefronts are generated, apparently originating at that edge. These secondary wavefronts will interfere with the primary wavefront as well as with each other to form various diffraction patterns.

diffraction grating

A glass substrate carrying a layer of deposited aluminum that has been pressure-ruled with a large number of fine equidistant grooves, using a diamond edge as a tool. Light falling on such a grating is dispersed into a series of spectra on both sides of the incident beam, the angular dispersion being inversely proportional to the line spacing. By proper shaping of the diamond edge, however, the grooves can be formed in such a way as to concentrate most of the energy into a single spectral order; such a grating is said to be blazed. Cast replica gratings can be made in plastic or another substrate, the replica being used either as a transmitting grating or by aluminizing it, as a reflecting grating. Plane gratings require external optics to focus the spectral lines, but a grating ruled on a concave surface is self-focusing.

diode

A two-electrode device with an anode and a cathode that passes current in only one direction. It may be designed as an electron tube or as a semiconductor device.

diode-pumped solid-state laser

A compact solid-state laser, referred to as DPSSL, created when a laser diode pumps light into either the sides or end of gain crystal. Depending on where the pumping occurs, high-quality and stable output beams can be achieved with end-pumped lasers, and high-power output beams can be realized as with side-pumped lasers. Common in green laser pointers, DPSSLs are also used for a variety of materi-

---

• For a comprehensive dictionary of optics and photonics terms, please visit photonics.com/EDU.

als processing applications, such as cutting and precision marking, and in optogenetics.

direct laser interference patterning
Also called DLIP, a high-speed, high-resolution processing technique that uses high-power, pulsed laser systems to directly ablate micro- and nanoperiodic structures with different features on large and geometrically varying surface areas, improving their friction, wear, light management, adhesiveness, biocompatibility and other physical and chemical properties.

dispersion
The separation of a beam into its various wavelength components. In an optical fiber, dispersion occurs because the differing wavelengths propagate at differing speeds. Also called chromatic dispersion.

divergence
1. In optics, the bending of rays away from each other. **2.** In lasers, the spreading of a laser beam with increased distance from the exit aperture. Also called beam spread. **3.** In a binocular instrument, the horizontal angular disparity between the two lines of sight.

dopant
The impurity added to a substance to produce desired properties in the substance.

Doppler shift
The magnitude, expressed in cycles per second, of the alteration of the wave frequency observed as a result of the Doppler effect.

E

elasto-optic effect
A change in the refractive index of an optical fiber caused by variation in the length of the fiber core in response to mechanical stress.

electro-optic effect
The change in the refractive index of a material under the influence of an electrical field.

electro-optic material
A material having refractive indices that can be altered by an applied electric field.

electro-optic modulator
A device that uses an applied electrical field to alter the polarization properties of light.

electro-optics
1. The branch of physics that deals with the use of electrical energy to create or manipulate light waves, generally by changing the refractive index of a light-propagating material; **2.** Collectively, the devices used to affect the intersection of electrical energy and light. Compare with optoelectronics.

ellipsometry
The measurement of the change in ellipticity of an optically polarized light beam after reflection from a surface in a particular manner.

emission spectrum
The spectrum formed by radiation from an emitting source, in contrast to absorption spectra.

emitter
A source of radiation.

energy density
The energy in a medium per unit volume.

excimer
A contraction of "excited dimer." The term refers to an excited species made by combination of two identical atoms or molecules, one of which is excited and one of which is at a ground state.

excimer laser
A rare-gas halide or rare-gas metal vapor laser emitting in the ultraviolet (126 to 558 nm) that operates on electronic transitions of molecules, up to that point diatomic, whose ground state is essentially repulsive. Excitation may be by E-beam or electric discharge. Lasing gases include ArCl, ArF, KrCl, KrF, XeCl and XeF.

excitation
1. The process by which an atom acquires energy sufficient to raise it to a quantum state higher than its ground state. **2.** More specifically with respect to lasers, the process by which the material in the laser cavity is stimulated by light or other means, so that atoms are converted to a semistable state, initiating the lasing process.

excited state
The stationary state of an ion, atom or molecule, above the ground state that is produced by the interaction with the radiation field or another ion, atom or molecule.

F

fan
A set of rays through a lens originating at a common point and contained in one plane.

Faraday Isolator
An optical device that acts as a "one way valve" for collimated light (laser beams), passing it in one direction, but not the other. It consists of three components, a linear polarizer, followed by a Faraday rotator (set to rotate the input

polarization by 45°), followed by another polarizer oriented at 45° to the first. Light traveling "forward" through the device is first linearly polarized; its polarization vector is then rotated 45° by the Faraday rotator which is then passed by the final polarizer. Traveling in the reverse direction, light is again linearly polarized and then rotated 45°. This orients the polarization at 90° to the first polarizer, which therefore blocks it. Faraday isolators are most typically used to prevent feedback into a laser cavity which improves stability.

femtosecond laser

A type of ultrafast laser that creates a minimal amount of heat-affected zones by having a pulse duration below the picosecond level, making the technology ideal for micromachining, medical device fabrication, scientific research, eye surgery and bioimaging.

fiber bundle

A rigid or flexible, concentrated assembly of glass or plastic fibers used to transmit optical images or light.

fiber laser

A laser in which the lasing medium is an optical fiber doped with low levels of rare-earth halides to make it capable of amplifying light. Output is tunable over a broad range and can be broadband. Laser diodes can be used for pumping because of the fiber laser's low threshold power, eliminating the need for cooling.

fiber optics

The use of thin flexible glass or plastic fibers as wave guides — or 'light pipes'— to channel light from one location to another. Fiber optics is based off of the principle of total internal reflection (TIR). Light enters the fiber at a particular angle of incidence. When this angle becomes larger than the 'critical angle' of the fiber, the light is then reflected at the surface of the fiber and no internal losses are encountered. The light is then 'trapped' and will continue to bounce back and forth at this angle down the fiber seeing very minimal loss along propagation. Fiber optics is the basis for optical communication systems.

filter

With respect to radiation, a device used to attenuate particular wavelengths or frequencies while passing others with relatively no change.

fluence

A measure of time-integrated particle flux given as particles per square centimeter (joules/cm^2).

flux

Time rate of flow of energy; the radiant or luminous power in a beam.

focal point

That point on the optical axis of a lens, to which an incident bundle of parallel light rays will converge.

frequency doubling

A nonlinear optical process in which the frequency of an optical beam is doubled coherently.

fringe

An interference band such as Newton's ring.

full width half maximum

Full width half maximum (FWHM) is a measure of the extent of a function. Given by the difference between the two extreme values of the independent variable at which the dependent variable is equal to half of its maximum value. The term duration is preferred over width when the independent variable is time. Commonly applied to the duration of pulse waveforms, the spectral extent of emission or absorption lines, and the angular or spatial extent of radiation patterns.

fused silica

Glass consisting of almost pure silicon dioxide (SiO_2). Also called vitreous silica. Frequently used in optical fibers and windows.

fusing

The permanent uniting of two glass pieces by high-temperature heating.

G

gain

Also known as amplification. **1.** The increase in a signal that is transmitted from one point to another through an amplifier. A material that exhibits gain rather than absorption, at certain frequencies for a signal passing through it, is known as an active medium. **2.** With reference to optical properties, the term may be defined in two ways: a. the relative brightness of a rear projection screen as compared with a perfect lambertian reflective diffuser; b. the ratio of brightness in footlamberts to incident illumination in footcandles. **3.** In a photodetector, the ratio of electron-hole pairs generated per incident photon.

gas discharge

The conduction of electricity in a gas as a result of the ions generated by collisions between electrons and gas molecules.

gas laser

One of the first lasers to find practical application. Generally, the pumping mechanism is an electric discharge, although some high-power forms employ chemical reaction or gas compression and expansion to form population inversion. Vibrational energy level transitions give emis-

• For a comprehensive dictionary of optics and photonics terms, please visit photonics.com/EDU.

sion from the near-infrared to far-infrared, and vibrational rotational transitions give emission from the far-infrared to microwave wavelengths. The helium-neon, argon-ion and carbon dioxide lasers are examples of gas lasers.

Gaussian beam
A beam of light whose electrical field amplitude distribution is Gaussian. When such a beam is circular in cross section, the amplitude is $E(r) = E(0) \exp[-(r/w)^2]$, where r is the distance from beam center and w is the radius at which the amplitude is 1/e of its value on the axis; w is called the beamwidth.

grating
A framework or latticework having an even arrangement of rods, or any other long narrow objects with interstices between them, used to disperse light or other radiation by interference between wave trains from the interstices. The ability of a grating to separate wavelengths (chromatic resolving power) is expressed as being equal to the number of lines in the grating.

H
halide
An MX-type compound of which fluorine, chlorine, iodine, bromine or astatine is a constituent. Glasses based on halides, in particular heavy metal fluoride glass (HMFG), demonstrate promise for infrared fiber transmission over very long distances.

HALT
highly accelerated life test

heat sink
A series of flanges or other conducting surfaces, usually metal, attached to an electronic device to transmit and dissipate heat that might damage internal circuitry.

heat treating
The process of subjecting glass to temperature cycling to produce physico-chemical reactions that alter its properties. Similar processes can be accomplished with a laser — most commonly a CO_2, Nd:YAG or Nd:glass laser — in certain kinds of metalworking including surface hardening, cladding, alloying and glazing (skin melting).

hot spot
Term applied to laser technology to denote an area of above-average intensity often attributable to atmospheric inconsistencies.

I
incidence
Flux incident per unit area of a surface.

index of refraction
The ratio of the velocity of light in a vacuum to the velocity of light in a refractive material for a given wavelength.

index profile
In an optical waveguide, the refractive index as a function of radius.

integrated laser
A type of laser for which a large number of the components can be fabricated in or upon a single substrate.

intensity
Flux per unit solid angle.

interference
1. The additive process whereby the amplitudes of two or more overlapping waves are systematically attenuated and reinforced. **2.** The process whereby a given wave is split into two or more waves by, for example, reflection and refraction of beamsplitters, and then possibly brought back together to form a single wave.

interferometer
An instrument that employs the interference of lightwaves to measure the accuracy of optical surfaces; it can measure a length in terms of the length of a wave of light by using interference phenomena based on the wave characteristics of light. Interferometers are used extensively for testing optical elements during manufacture. Typical designs include the Michelson, Twyman-Green and Fizeau interferometers. The basic interferometer components are a light source, a beamsplitter, a reference surface and a test surface. The beamsplitter creates the reference and test beams from a single light source.

internal
With reference to absorbance, absorptance, transmittance and the like, the processes occurring within a specimen between the entry and exit surfaces.

ion
An atom that has gained or lost one or more electrons and, as a result, carries a negative or positive charge.

ion laser
A laser in which the transition involved in stimulated emission of radiation takes place between two levels of an ionized gas. The gases are electrically excited in a container called a plasma tube, which typically consists of an alumina or ceramic envelope that is vacuum sealed at each end by either two Brewster windows or one Brewster window and one sealed-cavity mirror. The optical cavity is defined by a 100-percent-reflecting mirror and a partially transmissive output coupling mirror. It provides moderate to high

continuous-wave output of typically 1 mW to 10 W. For single-frequency operation, the high reflector is replaced with a Brewster prism, and an etalon is inserted.

irradiance

Radiant flux incident per unit area of a surface. Also called radiant flux density.

irradiation

Application of radiation to an object.

isolator

A device intended to prevent return reflections along a transmission path.

isotropic

That property of a material that determines that velocity of propagation within the material is the same for all directions.

K
kerf

The material lost during a laser cutting or machining operation.

Kerr effect

A quadratic nonlinear electro-optic effect found in particular liquids and crystals that are capable of advancing or retarding the phase of the induced ordinary ray relative to the extraordinary ray when an electric current is applied. It varies as the square of the voltage.

keyhole welding

The process of binding or attaching larger metal sheets by laser welding. The effect is generated by higher power densities which, while creating a larger weld, produce a vapor that is penetrated by the beam to produce an ideal effect.

kinematic mount

A mount for an optic element or optics assembly, designed so that all six degrees of freedom are singly constrained. This assures that movement will be prevented, while stress will not be introduced into the optics.

L
ladar

An acronym of laser detection and ranging, uses laser light for detection of speed, altitude, direction and range; it is often called laser radar.

laser

A device which operates under the processes of absorption and stimulated emission and by the condition that gain exceeds loss in order to sustain amplification. The term laser is the acronym for light amplification by the stimulated emission of radiation.

The standard components of a laser include: 1.) a lasing material known as the gain medium. 2.) a pump source. 3.) laser cavity.

To achieve lasing, the atoms of a material such as crystal, glass, liquid, dye or gas are excited by the pump source to a semistable state. The pump source is conventionally another light source (i.e. a laser diode or flash lamp) or an electric discharge. The light emitted by an atom as it drops back to the ground state interacts with nearby excited atoms to release identical pairs of photons in the process called stimulated emission. This process is duplicated as the photons bounce back and forth in the cavity from mirrors or other reflective cavity structures which then further amplify the light emission, producing beams of light at specific frequencies; traditionally, lasers comprise cavities in which a pair of mirrors (one highly reflective and one partially reflective) bounce light between them for amplification.

Lasers are used in a wide range of applications and industries including manufacturing, consumer electronics, medicine, sensing, aesthetics, basic scientific research, entertainment and more.

laser ablation

The removal of material from a surface by high intensity pulsed or CW laser radiation emission.

laser cavity

A means of optical confinement intended to increase the gain length of radiation prior to emission from the device. The means of optical confinement used to increase gain path length vary depending upon the properties of the beam desired within the lasing medium. High light intensities occur within a laser cavity and dielectric mirrors coated for the lasing wavelength are used. The position and curvature of the optical cavity elements may be altered in order to optimize the laser performance as needed for a particular application.

laser drill

High power laser ablation device that by pulsed operation produces holes of controllable dimension on the scale of microns. Laser drill applications include turbine manufacturing and surface metal processing.

laser fusion

Optical confinement of matter with high field energies intended to induce a stable nuclear fusion interaction.

laser head

Contains elements which produce lasing., e.g. gain medium, oscillator mirrors, as well as housing.

• For a comprehensive dictionary of optics and photonics terms, please visit photonics.com/EDU.

laser oscillator
Contains the light or beam path within a laser device. The oscillator uses reflective optical components that are oriented to obtain multiple confined reflections.

laser rod
In a solid-state laser, the material (Nd:YAG, Nd:glass, ruby) in which lasing action takes place.

laser tube
The device, usually made of glass or a similar material, that contains the resonant cavity and optics of a gas laser.

lasing medium
The material that produces stimulated emission from within a laser oscillator. Laser gain media may vary from extended-length glass fibers to submicron-length semiconductor material.

lens
A transparent optical component consisting of one or more pieces of optical glass with surfaces so curved (usually spherical) that they serve to converge or diverge the transmitted rays from an object, thus forming a real or virtual image of that object.

lidar
An acronym of light detection and ranging, describing systems that use a light beam in place of conventional microwave beams for atmospheric monitoring, tracking and detection functions.

light
Electromagnetic radiation detectable by the eye, ranging in wavelength from about 400 to 750 nm. In photonic applications light can be considered to cover the nonvisible portion of the spectrum which includes the ultraviolet and the infrared.

light source
The generic term applied to all sources of visible radiation from burning matter to ionized vapors and lasers, regardless of the degree of excitation.

linear amplifier
Amplifier in which the input and output pulse heights are directly proportional.

M
mask
1. A framelike structure that serves to restrict the viewing area of the screen when placed before a television picture tube. 2. In photolithography, a photomask (or mask) is typically a patterned transparent plate or an opaque plate with patterned holes or transparencies that uses a laser light source to transfer and print the pattern by an etching process onto a substrate that is typically a silicon wafer used in integrated circuitry.

member
In a lens system, a group of elements considered as an entity; either a front or rear member depending on whether it is before or after the aperture stop.

metrology
The science of measurement, particularly of lengths and angles.

microelectromechanical systems (MEMS)
Refers to micron-size complex machines that have physical dimensions suitable for the fabrication of optical switches for use in state-of-the-art communications networks.

microlithography
A technique for producing micron-size structures on surfaces by using short-wavelength light or electron beams.

micrometer
1. The SI unit of length equal to 10^{-6} m. Also called micron.
2. A screw thread device used to make accurate physical linear measurements.

minimum spot size
The smallest linear diameter to which a laser or other beam of radiant energy is capable of being focused, depending on the quality of the focusing optics, aperture and focal length, beam irradiance distribution (whether uniform or Gaussian), wavelength and other factors.

mirror
A smooth, highly polished surface, for reflecting light, that may be plane or curved if wanting to focus and or magnify the image formed by the mirror. The actual reflecting surface is usually a thin coating of silver or aluminum on glass.

mixing
Combining light beams, usually of unlike frequencies, to form a single beam with a frequency that is equal to the frequency sum or difference of the original beams. The resultant "beat" frequency is often predetermined for the given application.

mode
1. The characteristic of how light propagates through a waveguide that can be designated by a radiation pattern in a plane transverse to the direction of travel. 2. The state of an oscillating system such as a laser that corresponds to both a particular field pattern the system generates, as well as one of the possible resonant frequencies of the system. In a laser system, the field pattern produced as described above

is referred to as a transverse mode and takes the common name(s) Gaussian, Hermite Gaussian, Laguerre-gauss etc. depending on the transverse mode. The resonant frequencies described refer to a longitudinal mode which is the light wavelength at which the laser is designed to operate.

mode-locked laser

A laser that functions by inducing a fixed phase relationship between all of the modes present in the laser cavity. Once all modes are *in phase*, the modes are then said to be locked. The interference of these modes with one another inside the cavity then allows the laser to produce high energy, peak-powered short pulses of light in time durations that range from femtoseconds to picoseconds.

modulation

In general, changes in one oscillation signal caused by another, such as amplitude or frequency modulation in radio which can be done mechanically or intrinsically with another signal. In optics, the term generally is used as a synonym for contrast, particularly when applied to a series of parallel lines and spaces imaged by a lens, and is quantified by the equation: Modulation = (Imax – Imin)/(Imax + Imin) where Imax and Imin are the maximum and minimum intensity levels of the image.

modulator

See electro-optic modulator, optical modulator.

N

nanotechnology

The use of atoms, molecules and molecular-scale structures to enhance existing technology and develop new materials and devices. The goal of this technology is to manipulate atomic and molecular particles to create devices that are thousands of times smaller and faster than those of the current microtechnologies.

near-infrared (NIR)

The shortest wavelengths of the infrared region, nominally 0.75 to 3 µm.

near-ultraviolet

The longest wavelengths of the ultraviolet region, nominally 300 to 400 nm.

noise

The unwanted and unpredictable fluctuations that distort a received signal and hence tend to obscure the desired message. Noise disturbances, which may be generated in the devices of a communications system or which may enter the system from the outside, limit the range of the system and place requirements on the signal power necessary to ensure good reception.

normal

Sometimes referred to as the surface normal or "surface norm"; the normal is an axis that forms right angles with a surface that light is incident upon or with other lines. The normal is used to determine incident, reflective and refractive angles, as all of these angles are sketched and measured with respect to the normal of any given surface.

numerical aperture (NA)

The sine of the vertex angle of the largest cone of meridional rays that can enter or leave an optical system or element, multiplied by the refractive index of the medium in which the vertex of the cone is located. Generally measured with respect to an object or image point, and will vary as that point is moved. The numerical aperture of an optical system is critical in determining the resolution limits along with the diffraction limited spot size of a given optical system.

O

optical

Pertaining to optics and the phenomena of light.

optical component

One or more optical elements — typically cemented together — in an optical system that are treated as a single group; e.g., a beamsplitter, or a cemented doublet or triplet.

optical element

An optical part constructed of a single piece of optical material. It is usually a single lens, prism or mirror.

optical fiber

A thin filament of drawn or extruded glass or plastic having a central core and a cladding of lower index material to promote total internal reflection (TIR). It may be used singly to transmit pulsed optical signals (communications fiber) or in bundles to transmit light or images.

optical filter

A device with characteristics of selective transmittance, capable of passing a certain part of the electromagnetic spectrum while being opaque to the other portions. Means of producing filters varies considerably. Color filters (for the visual) usually consist of glass, gelatin or plastic containing dyes or pigments. Bandpass filters are found in signal processing and are commonly fabricated via electronics and complex circuitry and are designed to pass and reject parts of the spectrum based on the value of its frequency in comparison to the cutoff frequency. For a high pass optical filter, high-frequency signals are transmitted while frequencies lower than the cutoff frequency are rejected, and low pass filters are the opposite.

optical modulator

A multilayered thin-film device used to modulate transmitted light in integrated photonic circuits.

• For a comprehensive dictionary of optics and photonics terms, please visit photonics.com/EDU.

optical spectrum

1. Generally, the electromagnetic spectrum within the wavelength region extending from the vacuum ultraviolet at 40 nm to the far-infrared at 1 mm. **2.** The wavelength or color distribution found in a white light source once passed through a grating, prism, or other dispersive optical element.

optical surface

A reflecting or refracting surface contained within an optical system.

optical system

A group of lenses, or any combination of lenses, mirrors and prisms, so constructed as to refract or reflect light to perform some definite optical function.

optics

The study of light — optics is the area of physics that deals with the generation, propagation, and interaction of light waves in the electromagnetic spectrum with wavelengths (or frequencies) ranging from that of the ultraviolet to the deep infrared.

optoelectronics

A subfield of photonics that pertains to an electronic device that responds to optical power, emits or modifies optical radiation, or utilizes optical radiation for its internal operation. Any device that functions as an electrical-to-optical or optical-to-electrical transducer. Electro-optic often is used erroneously as a synonym.

output coupler

The partially reflective mirror at the end of the laser cavity that is the source of the beam. It controls the coupling percentage for high output power and maintains correct mode structure in the cavity.

P

peripheral

Near the boundary or edge of the field of an optical system; the outer fringe.

phase

In a periodic function or wave, the segment of the period that has elapsed, measured from some fixed origin. If the time for one period is expressed as 360° along a time axis, the phase position is called the phase angle.

phosphor

A chemical substance that exhibits fluorescence when excited by ultraviolet radiation, x-rays or an electron beam. The amount of visible light is proportional to the amount of excitation energy. If the fluorescence decays slowly after the exciting source is removed, the substance is said to be phosphorescent.

photolithography

A lithographic technique using an image produced by photography for printing on a print-nonprint, sectioned surface.

photon

A quantum of electromagnetic energy of a single mode; i.e., a single wavelength, direction and polarization. As a unit of energy, each photon equals hn, h being Planck's constant and n, the frequency of the propagating electromagnetic wave. The momentum of the photon in the direction of propagation is hn/c, c being the speed of light.

photonics

The technology of generating and harnessing light and other forms of radiant energy whose quantum unit is the photon. The science includes light emission, transmission, deflection, amplification and detection by optical components and instruments, lasers and other light sources, fiber optics, electro-optical instrumentation, related hardware and electronics, and sophisticated systems. The range of applications of photonics extends from energy generation to detection to communications and information processing.

photopolymer

A polymer produced as a result of photochemical processes.

photoresist

A chemical substance rendered insoluble by exposure to light. By means of a photoresist, a selected pattern can be imaged on a metal. The unexposed areas are washed away and are ready for etching by acid or doping to make a microcircuit.

photovoltaic cell

Also known as a self-generating barrier layer cell. A photoelectric detector that converts radiant flux directly into electrical current. Generally, it consists of a thin silver film on a semiconductor layer deposited on an iron substrate.

picosecond pulse

A pulse having extremely short duration, about 10^{-13} to 10^{-10} s, that is produced by mode locking of wide-bandwidth lasers, such as the organic dye and ruby lasers. Picosecond pulses are used in the study of extremely rapid decay processes.

pixel

Contraction of "picture element." A small element of a scene, often the smallest resolvable area, in which an average brightness value is determined and used to represent that portion of the scene. Pixels are arranged in a rectangular array to form a complete image.

plasma

A gas made up of electrons and ions.

plastic optics

The integration of plastic materials into optical applications. When the materials are refined into lenses, prisms and mirrors, they serve the purpose of glass optics at a lower cost and a significant savings of weight.

Pockels cell

A device containing an electro-optic crystal and using the Pockels effect. A voltage applied across the crystal generates birefringence, causing plane-polarized light propagating through the crystal to be resolved into two orthogonal vectors. The change in retardation between the two vectors (ellipticity) is proportional to the magnitude of the electrical field. A crossed polarizer analyzes the output beam, resulting in intensity modulation. Response time can be in picoseconds.

polar

Depicting one of the two ends of an axis of rotation.

polarization

With respect to light radiation, the restriction of the vibrations of the magnetic or electric field vector to a single plane. In a beam of electromagnetic radiation, the polarization direction is the direction of the electric field vector (with no distinction between positive and negative as the field oscillates back and forth). The polarization vector is always in the plane at right angles to the beam direction. Near some given stationary point in space the polarization direction in the beam can vary at random (unpolarized beam), can remain constant (plane-polarized beam), or can have two coherent plane-polarized elements whose polarization directions make a right angle. In the latter case, depending on the amplitude of the two waves and their relative phase, the combined electric vector traces out an ellipse and the wave is said to be elliptically polarized. Elliptical and plane polarizations can be converted into each other by means of birefringent optical systems.

polarizer

An optical device capable of transforming unpolarized or natural light into polarized light, usually by selective transmission of polarized rays.

power density

In laser welding or heat treating, the instantaneous laser beam power per unit area. This parameter is key in determining the fusion zone profile (area of base metal melted) on a workpiece.

power supply

Refers to the voltage and current necessary for the operation of circuit devices.

probe

Acronym for profile resolution obtained by excitation. In its simplest form, probe involves the overlap of two counter-propagating laser pulses of appropriate wavelength, such that one pulse selectively populates a given excited state of the species of interest while the other measures the increase in absorption due to the increase in the degree of excitation.

process control

The collection and analysis of data relevant to monitoring the rate and quality of industrial production, either continuously or in batches. Corrections can be made manually or automatically, via a feedback control loop.

proximity effect

The underexposure caused by the diffraction of light passing through small openings spaced closely together in masks used in photolithography.

pulse amplification

The compression and intensification of a laser pulse of a specific width into a smaller pulse width. A spherical cavity, in conjunction with a beam compressor, is efficient for pulse amplification. Cones and flats are highly effective when used in conjunction with swept-line foci.

pulse duration

The lifetime of a laser pulse, generally defined as the time interval between the halfpower points on the leading and trailing edges of the pulse.

pulse width

The interval of duration of a pulse.

pulsed laser

A laser that emits energy in a series of short bursts or pulses and that remains inactive between each burst or pulse. The frequency of the pulses is termed the pulse-repetition frequency.

Q
Q

The figure of merit of a resonator, defined as $(2p) \times$ (average energy stored in the resonator)/(energy dissipated per cycle). The higher the reflectivity of the surfaces of an optical resonator, the higher the Q and the less energy loss from the desired mode.

Q-switch

A device used to rapidly change the Q of an optical resonator. It is used in the optical resonator of a laser to prevent lasing action until a high level of inversion (optical

• For a comprehensive dictionary of optics and photonics terms, please visit photonics.com/EDU.

gain and energy storage) is achieved in the lasing medium. When the switch rapidly increases the Q of the cavity, a giant pulse is generated.

quantum
Smallest amount into which the energy of a wave can be divided. The quantum is proportional to the frequency of the wave. See photon.

quarter-wave plate
A plate made of a double-refracting crystal having such a density that a phase difference of one-quarter cycle is formed between the ordinary and extraordinary elements of light passing through.

quasi-CW laser
A laser that generates a succession of pulses at a high enough repetition rate to appear continuous. The pump source is switched on for short intervals, nominally equal to the lifetime of the population inversion of the gain medium, during the continuous operation of the laser. This results in a lower duty cycle which reduces heating and achieves higher output powers.

R

radiance
Radiant power per unit source area per unit solid angle. Usually it is expressed in watts/m^2/steradian.

radiation
The emission and/or propagation of energy through space or through a medium in the form of either waves or corpuscular emission.

radiation mode
A mode in an optical waveguide whose fields are transversely oscillatory everywhere external to the waveguide. It exists even in the limit of zero wavelength.

ray
A geometric representation of a light path through an optical device; a line normal to the wavefront indicating the direction of radiant energy flow.

reflectance
The ratio of reflected flux to incident flux. Unless otherwise specified, the total reflectance is meant; it is sometimes convenient to divide this into the sum of the specular and the diffuse reflectance.

reflection
Return of radiation by a surface, without change in wavelength. The reflection may be specular, from a smooth surface; diffuse, from a rough surface or from within the specimen; or mixed, a combination of the two.

reflector
A type of conducting surface or material used to reflect radiant energy.

refraction
The bending of oblique incident rays as they pass from a medium having one refractive index into a medium with a different refractive index.

regenerative amplifier
A type of multiple-pass amplifier in which no optical leakage is allowed until a finite number of passes has occurred; at this time the entire cavity output is released as one pulse.

remote laser welding
A robotic process commonly employed by automakers that enables high-speed and flexible production throughput by using swiveling optics for precise beam positioning.

remote sensing
Technique that utilizes electromagnetic energy to detect and quantify information about an object that is not in contact with the sensing apparatus.

repeatability
The degree to which a predetermined or previous setting of a positioning device can be duplicated by observance of the optical phenomena.

resonance
A large amount of vibration in a system due to a small periodic stimulus that has about the same period as the natural vibration period of the system.

resonator
A volume, bounded at least in part by highly reflecting surfaces, in which light of particularly discrete frequencies can set up standing wave modes of low loss. Often, in laser work, the resonator contains two facing mirrors that may either be flat (Fabry-Perot resonator) or have some spherical curvature, which together bind the lasing material that is referred to as the gain medium, and hence the optical cavity of a laser is where lasing occurs.

rise time
Measurement of the time elapsed during the current output change from 10 to 90 percent in a photoconductor.

S

saturable absorber
A laser dye whose absorption coefficient drops at high levels of incident radiation. The phenomenon is often called bleaching.

saturation

1. The decrease of the absorption (or gain) coefficient of a medium near some transition frequency when the power of the incident radiation near that frequency exceeds a certain value. As long as the absorption (or gain) coefficient is constant, the power absorbed (or emitted) by the medium is proportional to the incident power. However, there is always a limit to the rate at which the medium can absorb (or emit) power that is determined by the lifetimes of the energy levels involved. As this limit is reached, the induced transitions become rapid enough to affect the energy level populations, making them more nearly equal. **2.** With respect to color, attribute of a visual sensation that permits a judgment to be made of the proportion of pure chromatic (in contrast to achromatic) color in the total sensation. The psychosensorial correlate, or nearly so, of colorimetric purity.

scattering

Change of the spatial distribution of a beam of radiation when it interacts with a surface or a heterogeneous medium, in which process there is no change of wavelength of the radiation.

scribing

The process of perforating a silicon or ceramic substrate with a series of tiny holes along which it will break. Nd:YAG or CO_2 lasers are now routinely used.

semiconductor laser

A semiconductor material which is designed and grown for the efficient production of short wavelength stimulated emission through high gain as well as low internal losses. Materials with band gap energies which emit radiation efficiently within the desired wavelength region are used. Diodes may emit vertically in relation to the laser material junction in a VCSEL configuration or horizontally in an edge emitting configuration. Synonymous with laser diode. Sometimes used interchangeably with diode laser, which more properly refers to a laser system composed of a gain element and external elements such as lenses and a grating to select a single wavelength.

sensitivity

In a radiation detector, the ratio of the output to the input signal.

sensor

1. A generic term for detector. **2.** A complete optical/mechanical/electronic system that contains some form of radiation detector.

silicon dioxide

An abundant material found in the form of quartz and agate and as one of the major constituents of sand. The silicates of sodium, calcium and other metals can be readily fused, and on cooling do not crystallize, but instead form the familiar transparent material glass.

slab laser

Solid-state laser geometry in which the standard rod is replaced by a slab of laser material. Often called total-internal-reflection face-pumped laser (TIR-FPL).

slit

An aperture, usually rectangular in shape, with a large length-to-width ratio, and a fixed or adjustable shape through which radiation enters or leaves an instrument. The aperture is generally small as compared to the light source.

solid-state laser

A laser using a transparent substance (crystalline or glass) as the active medium, doped to provide the energy states necessary for lasing. The pumping mechanism is the radiation from a powerful light source, such as a flash lamp. The ruby and Nd:YAG lasers are solid-state lasers.

source

A physical source of radiation, as contrasted to illuminant.

spatial resolution

In a vision system, the linear dimensions (X and Y) of the field of view, as measured in the image plane, divided by the number of pixels in the X and Y dimensions of the system's imaging array or image digitizer, expressed in mils or inches per pixel.

spectral

Pertaining to or as a function of wavelength. Spectral quantities are evaluated at a single wavelength.

spectral bandwidth

The wavelength interval in which radiant intensity is at least 50 percent of the maximum spectral value.

spectral width

A measure of the wavelength extent of a spectrum.

spectrum

See optical spectrum; visible spectrum.

SPIN

Acronym for self-aligned polysilicon interconnect N-channel. A metal-gate process that uses aluminum for the metal-oxide semiconductor (MOS) gate electrode as well as for signal and power supply connectors.

splitter

A passive fiber optic coupler that divides light from a single fiber into two or more fiber channels.

• For a comprehensive dictionary of optics and photonics terms, please visit photonics.com/EDU.

spontaneous emission

Radiation emitted when a quantum mechanical system drops spontaneously from an excited level to a lower level. This radiation is emitted according to the laws of probability without regard to the simultaneous presence of similar radiation. The rate of spontaneous emission is proportional to the Einstein "A" coefficient and is inversely proportional to the radiative lifetime.

standing wave

The combination of two waves having the same frequency and amplitude and traveling in opposite directions. Standing waves are indicated by a stationary set of nodes spaced one-half wavelength apart along the propagation direction of the waves.

stereolithography

A method of creating real three-dimensional models by using lasers driven by CAD software. In contrast to the normal practice of removing material, this process polymerizes a liquid to quickly produce shapes that are untouched by human hands or cutting tools. Also known as three-dimensional imaging and three-dimensional modeling.

stripper

A tool used to remove the outer cladding of an optical fiber without damaging the fiber core.

sync

Abbreviation of "synchronization." In television, the timing signals used to drive the scanning process. Horizontal sync triggers the retracing of the raster line beginning at the left-hand side of the screen; vertical sync triggers the beginning of a new field.

system

A combination of components arranged so as to perform at least one function.

T

tap

A device for extracting a portion of the optical signal from a fiber.

termination

The process of attaching a device to the end of a fiber cable to enable the beam to be transmitted farther, to pass to another device and to avoid reflection.

thin film

A thin layer of a substance deposited on an insulating base in a vacuum by a microelectronic process. Thin films are most commonly used for antireflection, achromatic beamsplitters, color filters, narrow passband filters, semitransparent mirrors, heat control filters, high reflectivity mirrors, polarizers and reflection filters.

threshold

1. In visual perception, the minimum value of stimulus that can be perceived on the average. **2.** In optical detection systems, that signal level at which the probability of detection is 50 percent.

translation stage

A small optical mounting platform that is usually mobile on ball bearings and can be driven manually or by motor to provide precision movement along two axes.

two-photon polymerization

An additive fabrication technique, referred to as TPP, used to make 3D microstructures with submicron feature sizes by using a near-infrared (NIR) emission that excites a photosensitive resin, triggering multiphoton absorption where light intensity is highest and a polymerization process that changes it from a liquid to a solid. When the volume of the focused laser beams, or voxels, are precisely overlapped, 3D microstructures are created and revealed by washing away unsolidified resin with an organic solvent.

U

ultrashort-pulse laser

A laser capable of generating light pulses that last only a few femtoseconds. This can be achieved by nonlinear filtering to increase bandwidth and compress the pulse or by passive mode-locking or synchronous pumping in conjunction with pulse-shaping techniques.

ultraviolet A

The region of the electromagnetic spectrum from 320 to 400 nm.

V

V-groove

A V-shaped channel pressed or etched into a substrate, in which, for example, optical fibers may be placed to create an integrated optical component.

vergence

The angular relation between two light rays that originated at the same object point. Sometimes used to indicate the angle between the visual axes of the eyes.

via

In integrated circuits, a pathway, hole or other passage through the substrate.

visible spectrum

That region of the electromagnetic spectrum to which the retina is sensitive and by which the eye sees. It extends from about 400 to 750 nm in wavelength.

voxel

An element within a three-dimensional data set image.